BY THE SAME AUTHOR

Rainbows, Snowflakes, and Quarks

TAMING THE ATOM

HANS CHRISTIAN VON BAEYER

TAMING THE ATOM

The Emergence of the Visible Microworld

VIKING

VIKING

Published by the Penguin Group
Penguin Books Ltd, 27 Wrights Lane, London W8 5TZ, England
Penguin Books USA Inc., 375 Hudson Street, New York, New York 10014, USA
Penguin Books Australia Ltd, Ringwood, Victoria, Australia
Penguin Books Canada Ltd, 10 Alcorn Avenue, Toronto, Ontario, Canada M4V 3B2
Penguin Books (NZ) Ltd, 182–190 Wairau Road, Auckland 10, New Zealand

Penguin Books Ltd, Registered Offices: Harmondsworth, Middlesex, England

First published in the USA by Random House Inc. 1992
First published in Great Britain by Viking 1993
1 3 5 7 9 10 8 6 4 2

Printed in England by Clays Ltd, St Ives plc

A CIP catalogue record for this book is available from the British Library

ISBN 0-670-84017-3

For Madelynn and Lili

"What is REAL?" asked the Rabbit one day, when they were lying side by side near the nursery fender, before Nana came to tidy up the room. "Does it mean having things that buzz inside you and a stick-out handle?"

"Real isn't how you are made," said the Skin Horse. "It's a thing that happens to you."

—Margery Williams, *The Velveteen Rabbit*

Acknowledgments

I would like to thank the editors of *The Sciences* and of *Discover* for allowing me to use material that appeared in preliminary form in the pages of their magazines. My research was made pleasurable by the generous help of many colleagues, and especially by the hospitality of David Wineland, Sam Hurst, Jean-Pierre Vigier, and Ken Snelson, and I am also deeply grateful to my agent, Beth Vesel, on whose calm, confident advice I have come to rely in times of stress. This book is dedicated to my children, who have waited patiently for two years to play "battleships" with me again, but without my wife Barbara Watkinson it would be just paper.

Contents

Prologue

The first time I saw an atom, it blinked—it took me by surprise, winking like some living thing. Ever since I first learned about atoms as a boy, I had become accustomed to thinking of them as microscopic bits of inert, inanimate matter that could do no more than react to external stimuli, like grains of sand swept along the beach by wind and waves. But this atom, which happened to be one of mercury, was blinking by itself as if in response to some internal influence. I felt that I had been granted an unexpected, fleeting glimpse beyond its outward appearance into the mysterious hidden world below the atomic surface.

The art of capturing and isolating atomic particles is new. The first photograph of an individual atom, taken at the University of Heidelberg in Germany, was published in 1980, and a decade later about half a dozen laboratories around the world had duplicated the feat. As a theoretical physicist I have used mathematics to describe atoms throughout my career, but the notion of catching them in action excited my curiosity. So I called around to learn more about it and was delighted when David Wineland, the scientist responsible for this type of work at the National Institute of Standards and Technology in Colorado, invited me to come out for a visit. Although I was familiar with reproductions of his atomic images in the technical literature, I couldn't resist the opportunity to see an atom with my own eyes. So I packed my overnight bag and booked a flight to Boulder.

The National Institute of Standards and Technology occupies a cluster of low, gray buildings that seem to grow out of the red cliffs in the foothills of the Rockies like some alien outcropping. Its rooftop antennas and microwave dishes have a tidier, more professional look than ordinary satellite TV receivers and thus betray the scientific mission of the place. After crossing a marbled lobby the visitor ascends an unexpected flight of steps (presumably to clear a rise in the rocky terrain) and enters a long, dim corridor. It is decorated, like corridors of scientific laboratories everywhere, with posters of Albert Einstein sticking out his tongue, clippings of cartoons about crazed scientists in white coats, announcements of topical conferences in exotic parts of the world, graphs documenting the latest local achievements, and a collection of sundry bits of apparatus. The strangest object here is a grandfather clock made of transparent plastic, a reminder that the labs on both sides of the hallway belong to the Institute's Time and Frequency Division, the nation's official timekeeper. The chief justification for David Wineland's work is the search for the perfect clock, for which a solitary, trapped atom is a good candidate.

David, tall, fit, taciturn, and in his forties, did not wear a white coat, but jeans and a sweater. (Single atoms don't leave stains.) The door of his lab bore a warning to beware of the powerful laser light within, and the room itself, no larger than an average living room, was packed with equipment, the most prominent of which were three optical tables—steel behemoths designed to inhibit vibrations by sheer weight, whose surfaces are polished to an uncommon degree of flatness to facilitate delicate alignments.

The day I visited, all three were bristling with optical gadgetry. At first glance the lenses, mirrors, adjustable irises, prisms, baffles, and filters appeared to be scattered about at random, but the thin filaments of laser light that connected them in a delicate web of multicolored lace bespoke an intricate overall design. Some light beams threaded back and forth among the tables, tying the whole room into a single integrated instrument, and I had the uneasy feeling that if I were to nudge something accidentally, the whole experiment would be ruined. Only a cheap flashlight left incongruously in the middle of one of the tables relieved the look of untouchable perfection.

Behind the tables stood two enormous wooden crates with mysterious protruding tubes and wires, and while I was wondering which one might be the atomic trap, David opened a drawer and pulled out a little plastic box that contained a penny. On it, barely covering up the words E PLURIBUS UNUM, he had glued, for the benefit of visitors like me, one of his traps. It consisted of a miniature doughnut-shaped ring and what looked like pistons on either side of it, loosely covering the hole in the doughnut, but not quite touching it. All three pieces—the ring and the two end-caps—were made of flinty gray molybdenum metal that had been machined with exquisite precision. I was taken aback by the trap's diminutive size, but realized that from the microscopic vantage point of an atom, the open space inside the ring, between the two end-caps, represented a vast cavern like the Superdome. In use the trap would be electrically charged in such a way that its walls would repel the atom, which itself would also be charged by the loss of one of its electrons, and thus be forced to jiggle around in the center of the cavity.

The trap that was in actual operation was suspended inside a glass vacuum vessel in one corner of the lab. A good vacuum is essential for isolating atoms, otherwise they get jostled about and are eventually pushed out of the trap by a sea of air molecules. Mounted above the trap was an ultraviolet light detector connected to a monitoring screen. David explained that he would show me a mercury atom illuminated by ultraviolet light, which is invisible to the unaided eye. The pretty ribbons of laser light visible throughout the laboratory, it turned out, merely served a variety of secondary purposes, such as the generation and calibration of the primary, invisible beam.

The fact that I would not be seeing the atom directly but through so much intricate machinery didn't really bother me. Instruments for enhancing visual acuity are common—whether they are as simple as a pair of glasses or as sophisticated as the image intensifiers that capture color pictures of distant galaxies, the distinction is a matter of degree, not principle. The interposition of some amplifying device, such as an ultraviolet light detector, between me and my atom, would not, I thought, detract from the experience of seeing it. Just so I could be sure I knew what I was looking at.

David then pointed out a pencil-thin glass tube at the back of the vacuum apparatus containing a few drops of mercury, which served as the source of atoms to be trapped. He explained that it wasn't ordinary mercury of the kind used in barometers and thermometers. Mercury found in nature consists of a mixture of several different kinds of atoms that look the same and are chemically identical but have slightly different weights, just as grains of sand do. Precise control over weight is crucial for the atom trapper, so David had had to search for a source of mercury of unsurpassed purity that had even been sorted by atomic weight. He had been pleasantly surprised to find it at his own agency: the National Bureau of Standards, venerable predecessor of the National Institute of Standards and Technology, had sequestered a minuscule quantity of artificial mercury in some forgotten safe, synthesized just after World War II by transmutation of another element by radiation from a nuclear reactor. In an ironic reversal of the dreams of the medieval alchemists, the original element had been gold. So the atom I was about to see was really, and appropriately, the noblest of them all—gold—which had been debased by modern sorcery into common mercury.

Before turning on the illumination, David let me squint through a window no larger than the peephole in my front door in order to see the interior of the vacuum vessel. With the aid of a miniature telescope I could make out the trap suspended in space by a wire. The narrow gap between the ring and one of the two end-caps turned out to be a convenient keyhole for observing what went on inside. The ultraviolet detector, David said, was focused through that gap on a spot in the middle of the trap, where the atom was caught even as I was looking at it, though of course it was too small to be seen in the dim light inside the vessel.

Eventually David switched on the monitor. My first impression of what appeared was of snow falling in brilliant sunshine. Each twinkle represented a single particle of light, a photon, shot at the trap by the ultraviolet laser, and reflected off the molybdenum surface. Since I had by now seen the little doughnut-shaped ring twice, first on the penny and then in the actual trap itself, I had no trouble recognizing its outline on the screen. But the space in the middle was dark. David explained that he would have to fine-tune the frequency of the ultraviolet laser to the value best suited for showing mercury atoms. So he proceeded to do that, and then, finally, I saw my first atom.

Right in the middle of the trap a little star appeared. Tentatively at first, amid the flickering reflections all around, and then with increasing intensity, the mercury atom poured out its light. Held in a tight grip by electrical forces between its own charge and the metal walls of the trap, it did not budge; its trembling motion in the trap was far too minute to be noticeable. It looked firmly anchored, and indeed it was. David told me that other atoms had been kept in place in this same manner for as long as ten days, before some chance event, such as a collision with a stray molecule of air or a glitch in the electrical supply, caused them to drift into oblivion. So here it was, an atom in captivity.

It was then, as I watched spellbound, that I began to notice that the atom was blinking. At first I thought that this was just part of the general flickering of the screen, but it soon became apparent that the mercury atom was definitely turning off and on, at the rate of several times a second. This was surely the most astonishing thing I had ever seen. Whatever might be the cause of this phenomenon, it was a powerful reminder that atoms are active, dynamic systems, capable of the most intricate internal transformations and convolutions, and not in the least bit like the immutable, eternal kernels of matter the ancients had imagined them to be. Although I understood this difference intellectually, it took the impudent winking of a trapped mercury atom to drive the point home to me in an unforgettable way.

David emphasized that the blinking was far more than an unwelcome distraction. It was, in fact, the best proof that we were indeed looking at a single atom, and not, say, a little droplet of mercury, consisting of billions of atoms. Such a globule would also reflect ultraviolet light, but, unlike the atom, it would do so in a steady, uninterrupted beam. The size of the bright dot on the monitor would not have helped to distinguish an atom from a droplet—it is much larger than either one—but the blinking did; it turned out to be a crucial part of the experiment, a signature of a single atom.

Individual atoms reflect light differently from the way mirrors, or droplets of mercury, do; they first absorb and almost immediately reradiate particles of light. Absorption can only occur when the atomic electrons are in a specific configuration, or quantum state. The atom has to be in a receptive mode, so to speak. Sometimes, however, spontaneously and without any external stimulus, its electrons rearrange themselves into different configurations. They hop about inside the atom, executing what are called quantum jumps, and when they happen to end up in an unreceptive state, the atom is no longer able to absorb light: it blinks off. As soon as it jumps back to its original state, it blinks on again. This phenomenon was first suggested as a theoretical possibility by Niels Bohr in 1913, but until single atoms were caught in traps, it had not been observed experimentally. So in the end David delivered more than he had promised me on the phone. He showed me not only an atom but also a hint of its true quantum mechanical nature.

As I watched the mercury atom that day, I began to understand the fascination of trapping experiments. Even during the few minutes that I observed it, and started to make out the pattern of its blinking, I could feel the beginning of a process of familiarization that Antoine de Saint-Exupéry, in *The Little Prince,* called taming. "Taming is a thing that is too often neglected. It means 'to establish bonds,' " says the fox, one of the characters of that delightful fable. The course of acquaintance is slow: only after we become used to a thing can we comprehend it. The little prince understood the value of patience, for on his asteroid there grew a single rose that he had carefully tended and nurtured—that he had tamed.

We have all experienced the way in which taming leads to attachment. We have our favorite slippers, coffee mugs, and easy chairs, and we endow these objects with a significance that the bustling world around us cannot measure—we feel fondly toward them. We tame people as well: familiarity helps turn strangers into acquaintances, acquaintances into friends, and one particular friend into family (from which derives the word *familiar*). By taming atoms we can make them, too, familiar and can begin to understand them in a visceral way.

The pale spot of light in the center of the screen, amid the myriad random scintillations around it, reminded me of another picture. Toward the end of February 1990 the unmanned spacecraft Voyager 1 found itself at the edge of the solar system, thirteen years and three and a half billion miles from home, high above the plane of Earth's orbit. At the urging, in part, of the astronomer Carl Sagan, the little craft's camera was turned around for one last look at Earth, tiny among the stars and almost lost in the glare of our sun. The photo is not very revealing from a scientific point of view, but emotionally it packs a wallop, as Sagan well knew it would. It is the most distant picture of our planet ever taken, and it shows our insignificance in the vastness of the cosmos with unprecedented poignancy.

Earth, in this picture, is but a pale blue spot in a field of bright stars. A stray reflection in the camera casts a wide beam of light across the photograph, creating an effect that looks uncannily like the image on David Wineland's monitor, with its shimmering points of light and its reflections from the trap.

The two images represent the extremes of an enormous range of distances. Measured in powers of ten, or orders of magnitude, we are approximately as far removed from the one as from the other. Voyager 1, at the time of the photo, was about 10^{12} (a million million) meters away, while the atom is roughly 10^{-10} (one ten-billionth) meters across. Outside of that range, beyond the solar system, our telescopes show stars, galaxies, clusters and superclusters of galaxies, out to the visible horizon of the universe, but neither we nor our robotic ambassadors have ever been there. At the other extreme, inside the atom, accelerators reveal nuclei, and elementary particles, down to the elusive quarks and leptons that form the ultimate building blocks of matter, but we cannot see them and may never be able to.

The region between the two pictures—the accessible part of the universe—represents a large portion of the whole. The scale of distances from the diameter of the universe to the size of the smallest particle spans about forty-four orders of magnitude. Of this, the accessible scale, from 10^{12} to 10^{-10}, covers twenty-two orders of magnitude, or, in this way of reckoning, half of God's creation.

Much has been written in recent years about cosmology, the realm of the unimaginably large, and elementary-particle physics, the world of the indescribably small, and how the two meet in the Big Bang. While those subjects are intellectually fascinating, their unimaginability and indescribability render them highly abstract. The accessible universe, on the other hand, is visible and tangible. To people whose intuition has not been trained by the interpretation of scientific evidence this portion seems more real. And precisely because it is closer to our everyday life its mysteries are at least as compelling as those of the inaccessible world beyond.

The similarity between the two images at the edges of our direct experience is underscored by the fact that they were both taken in the same direction, as it were. One often sees juxtapositions of images of the macroworld with those of the microworld—a swirling galaxy, say, that resembles a gelatinous amoeba. But those pictures are made in opposite directions, both looking out from our own scale. The Voyager picture, on the other hand, followed by David Wineland's view of a mercury atom, represents a continuous zooming in over a vast scale, and for that reason the comparison is especially suggestive.

With respect to the public's understanding of science, however, the two pictures occupy different positions. While atomic physics is regarded as arcane knowledge reserved for a few chosen initiates, planetary astronomy has managed to catch the popular imagination. When I was a boy, we learned the names of the planets, from Mercury, which is next to the Sun, on out to Pluto, but those were just words. The planets themselves were remote and unreachable, mere dots in the sky. They were abstractions, in the original sense of the word, signifying "pulled away from" common experience. Yet with the space program the planets have become real; each one has acquired its own, distinctive personality. The forbidding craters of Venus, the red dirt of Mars, the great surging masses of methane on Jupiter, the icy etchings of the rings of Saturn, and the unforgettable blue-marbled surface of Earth itself are the stuff of popular news media. The planets, by becoming accessible, have become familiar. They have been tamed.

What has happened to the planets is about to happen to atoms. Within the decade of the 1980s no fewer than five Nobel Prizes were awarded for the development of techniques for manipulating and imaging individual atoms. As we are reaching down to the atomic level and making it accessible to our senses, we are taming it. Individual atoms can now be counted, and photographed, and kept in captivity; the surface roughness of materials can be magnified a millionfold to reveal its atomic character; and atoms can be combined one by one in the construction of synthetic materials. Soon the new images of atoms will become as familiar as those of the planets. The differences between atoms of mercury, gold, oxygen, and carbon will become manifest, and the elements will acquire distinctive personalities. Just as our appreciation of planet Earth has been transformed by our awareness of the findings of the space program, our relationship to the ordinary world around us will be changed by the opening of our eyes to the lavish beauty of the atomic landscape beneath its surface.

But seeing atoms is one thing, understanding their composition another. Just as we know that the pale blue dot in the Voyager photo is, in fact, a world of teeming life, we know that the mercury atom that looks like a flashing dot in the darkness of the vacuum is really a structure of wondrous complexity. Physicists have unraveled this structure down to its finest details and have learned to describe it with precision and confidence. Only the terms they use to talk about the interior of the atom are not the familiar words that describe our sense perceptions. Instead they speak in the altogether strange language of quantum mechanics, which describes atoms not as miniature objects like grains of sand but as immaterial clouds whose reality is questionable at best.

One of the architects of the quantum theory, Werner Heisenberg, came close to denying their reality altogether: "In the experiments about atomic events we have to do with things and facts, with phenomena which are just as real as any phenomena in daily life. But the atoms or elementary particles are not as real; they form a world of potentialities and possibilities rather than one of things or facts." Inside the atom reality seems to dissolve into enigma.

Until recently the inaccessibility of the atom had placed a barrier between the abstract world of quantum mechanics and what we experience as reality. Although the theory furnishes accurate predictions of experimental observations, the actual objects that quantum mechanics deals with, the individual atoms, were mere theoretical abstractions. Only large accumulations of them were observable, which led many scientists to believe that somehow the unsettling mysteries of quantum mechanics would be safely hidden under sheer numbers, that just as the law of averages produces order out of randomness, macroscopic reality emerges out of a collection of uncountable, unexplainable individual quantum events. But that hope has now vanished. Atoms appear before our eyes naked and single, and blink to remind us of their irreducible quantum mechanical nature.

Atoms straddle the boundary between what we see and what we know. Playwright Tom Stoppard neatly captured the fascination of that boundary in his spy drama, *Hapgood:* "There is a straight ladder from the atom to the grain of sand, and the real mystery is the missing rung. Above it, classical physics. Below it, quantum physics. But in between, metaphysics."

Metaphysics is a subject with which physicists do not meddle, for it is outside of their science, and as far as they are concerned, incomprehensible. Stoppard's epigram expresses the deeply troubling realization that there is an unbridged gulf between the world of our senses and the theoretical world that scientists have artificially constructed. The idea is puzzling, for if physics is an empirical science that proceeds from experience to abstraction, how could such a gap develop? Is this rift in our physical world picture something new, a consequence of some recent revolution in science, or is it so old that we have grown accustomed to it? How did we get into this alarming quandary?

The roots of the problem go back to the beginning of physics, when the Greek philosophers first wondered about the nature of the material world and created the theory of atoms. From the outset there was a gulf between theory and experience, and in succeeding centuries atoms sometimes seemed so remote that the theory lost credibility and faced oblivion. But again and again, in spite of the lack of direct evidence, atomism rebounded, until finally, in the early twentieth century, physicists began to experience atoms as real objects—only to have quantum mechanics open the divide between theory and common sense all over again.

Is this gap an unavoidable feature of the modern world picture, or will experiments like the trapping of atoms help to close it? The intellectual adventure story of the taming of the atom that ends at the forefront of modern research began more than two thousand years ago on the shores of the Aegean Sea.

The Past

1

The Enduring Idea of Atomism

"I had rather discover one true explanation than be king of Persia," the Greek philosopher Democritus once claimed. Today, two and a half millennia after the Persian empire and its Greek adversaries have long since perished, the atomic theory, Democritus's greatest creation, reigns as the sovereign paradigm of physical science. No empire can rival this idea in power and longevity. Democritus, in his wisdom, would have gotten the better part of the deal if it had been offered to him.

In view of his disdain for the trappings of worldly success, a conviction that determined the course of his life, it is ironic that Democritus is honored today by his portrait on a coin, where normally royalty and heads of state are pictured. The Greek ten-drachma piece depicts him in profile on one side of the coin, with an atom on the other. The philosopher's head is robust and square, set on a thick wrestler's neck. His massive skull is covered by a cap of short, curly hair that mingles with a beard of similar consistency. The face is flat, the nose gently curved, the jutting jaw set and determined. By raising the cheekbones and modeling the prominent eyebrows, the sculptor of this coin has managed to chisel out eyes that are ablaze with a ferocious intensity. The image conveys virility and power; the context reveals that the power is as much intellectual as physical.

But one detail of the otherwise lifelike portrait seems incongruous. The wrinkled forehead and the drawn corners of the mouth suggest a worried frown that belies Democritus's traditional epithet of "the laughing philosopher." Whether that legendary laughter was happy or sardonic, or, what is more likely, ironic, is open to debate, but we do know from his own writings that Democritus based his system of ethics on the idea that cheerfulness is a prerequisite to the good life. So it's a puzzle that on the ten-drachma coin he looks grim, not cheerful.

The solution appears on the opposite side of the coin: The atom pictured there is neither the primitive particle of Greek philosophy nor the enigmatic cloud of modern physics, but, for the sake of iconographic effectiveness, a version of Bohr's obsolete planetary model. The central nucleus is represented by a sunburst, which is of necessity far too large in proportion to the size of the atom (otherwise it wouldn't be visible), and is surrounded by three overlapping, circular electron orbits, seen from an oblique perspective that makes them appear elliptical. The electrons themselves are shown as dots, one on each path, and inasmuch as points have no internal parts, Democritus would have recognized *them* as atoms, rather than the entire planetary model. The three electrons are characteristic of an atom of lithium, the element that has been found to be essential for the maintenance of emotional equilibrium and the attainment of cheerfulness. Thus the atom provides a symbolic antidote for the philosopher's frown on the opposite side of the coin.

Between extensive world travels the real Democritus lived modestly in the Thracian town of Abdera, on the northern coast of the Aegean Sea, where modern Greece borders on Bulgaria. We can imagine him, a vigorous and lusty young man with a booming voice and a hearty appetite, in tireless discourse with his teacher, Leucippus. Why the ancient Greek philosophers, in defiance of universal human behavior, chose to cast aside greed, ambition, superstition, and aggressive impulses to devote themselves to strenuous inquiry and dispassionate debate about the nature of man and his world is a mystery, but their thinking shaped the development of Western civilization.

The particular question that preoccupied Leucippus and Democritus had been posed a century earlier by the legendary founder of philosophy, Thales of Miletus. "What is the nature of matter?" Thales had asked, and for "nature" he had used the word *physis,* which is the root of our word *physics.* The question remains forever fresh. It is asked today by six-year-old children—"What am I made of?"—and by physicists with billion-dollar research instruments, for it is as ancient as philosophy, as universal as curiosity, and as fundamental as language.

The atomic hypothesis, that matter consists of atoms, which was conceived by Leucippus and elaborated into a full theory by Democritus, is one answer to Thales' question. It originated as an ingenious compromise between two conflicting philosophical positions, the doctrines of the One and of the Many. The goal of mysticism, and one of the roots of religion, is the human desire to simplify the world by a belief in some universal unity. The philosophical search for the One led to attempts to explain matter in terms of a singular fundamental substance, or at least a small handful of substances, whatever they may be. Fire, water, air, and earth were early favorites in Greek and some Indian philosophical systems. However, the senses incontestably register a world of Many, of unfathomable complexity and boundless multiplicity. Related to this One-Many duality is the conflict between the search for permanent causes, for eternal verities, and the world's obvious state of continuous flux. Oneness versus multiplicity and persistence versus change are twin problems. How can the demands for simplicity and permanence be reconciled with the unfolding complexity that we see around us?

A clue came from the writings of an earlier thinker, the philosopher and admiral Melissus of Samos: "But if there are Many, they must each have the character of the One." This oracular epigram inspired the imaginative, but more concrete, suggestion of Leucippus and Democritus that if the world is made of innumerable individual atoms, which are identical to each other, then oneness resides in their constitution, and complexity in the infinite multiplicity of their arrangements. Furthermore, if they are indestructible, then permanence characterizes each atom, and mutability their spatial relationships. Persistence and change thereby become compatible. With one stroke pure reason had uncovered a fundamental truth about the nature of matter, and two seemingly contradictory views of the world were reconciled. A portrait on a coin of low denomination might seem an inadequate tribute to this monumental triumph of human intellect, but then again that portrait, like the atom itself, circulates in innumerable identical copies.

The essence of the theory, as stated by Democritus, is this:

> By convention there is sweet, by convention there is bitterness, by convention hot and cold, by convention color; but in reality there are only atoms and the void.

This formulation leaves no doubt as to what is real and what is mere appearance. People may well disagree about what they call bitter or sweet and hot or cold, but they would have to agree on the presence or absence of the small, solid particles of matter that are called atoms, if only the senses were sharp enough to detect them. The existence of the void, a hypothetical state of emptiness we now call the vacuum, was to be debated for the next two thousand years, but for Democritus it was merely a necessary corollary to the theory that matter is not continuous but granular—a property in which he found the ultimate reality.

How Leucippus and Democritus came to the atomic hypothesis is not known. The sources are fragmentary and later commentaries unreliable. Since Democritus was an outstanding mathematician, with particular strength in geometry, it is possible that he came upon the theory while searching for solutions to the mathematical paradoxes of his day. An example of such a dilemma concerned finding the formula for the volume of a cone. Democritus is believed to have discovered the simple but far from obvious recipe: Multiply one third of the height by the area of the base. It may be that the creation of the atomic hypothesis was prompted by the way he solved this mathematical problem.

Democritus thought of a cone as a solid object composed of a pile of infinitesimally thin, circular layers of material, like a stack of flimsy pancakes. The sizes of the pancakes puzzled him: If the solid figure were a cylinder instead of a cone, there would be no problem, since the layers would all be identical. But for a cone each layer must be slightly smaller than the one below it, so that the stack is tapered toward the top. When the edges of the pancakes are examined closely, however, the sides of the cone look like an ascending staircase, not like the smooth ramp that characterizes an idealized, mathematical cone.

This mathematical paradox this problem remained unsolved until the invention of calculus, two thousand years later, but as a physicist and believer in the existence of atoms, Democritus decided that each pancake represents one layer of atoms and that each is smaller than the one below it. Since atoms are too minute to be seen, the sides appear smooth to the senses, so there is no contradiction in the case of real, physical cones.

Inspired by such arguments, and guided by the ideas of his teacher Leucippus, Democritus managed by means of thought and intuition to "lift a corner of the great veil," as Albert Einstein said in another context, referring to the veil of mystery under which nature hides her operations. But science is a means, not an end, and for every explanation that unifies our understanding of the world we find a further mystery, a more basic question. For every mountain peak we conquer, there is a higher one in the distance from which we could see more, if we could only reach it.

Democritus answered Thales' question about the nature of matter by introducing a theory of invisible atoms, which in turn leads to the question What is the nature of those atoms? And when that question was answered in our own time, another price had to be paid; a new central mystery, the mystery of the quantum, then took the place of the enigma of matter and the riddle of the atom. And so it will go on, for nature is vast, and the veil opaque.

Yet even as an intermediate stage in the unending quest of science, the atomic theory of Democritus was by no means universally accepted. After a century of legitimacy as one of several rival philosophies, atomism ran afoul of the Philosopher himself, the venerable Aristotle, who gave it considerable attention in his commentaries, but in the end felt compelled to reject it. His verdict was that it was implausible because of its remoteness from the actualities of the world as perceived by the senses. Since Aristotle dominated the physical sciences for more than a millennium, almost to the exclusion of anyone else, atomism fell into decline; only a few dared to challenge the authority of the immortal sage of antiquity.

But the atomic doctrine was too powerful to die. During the time of Julius Caesar the Roman poet Lucretius became for all time its most eloquent exponent. His poem *De Rerum Natura,* translated sometimes as "On the Nature of Things"—or better, "On Nature"—is unique in that it functions simultaneously as poetry, scientific treatise, and moral tract, and succeeds spectacularly on all three levels.

As scientific text the poem derives its strength from the persistence with which it pursues its theme. After introducing the atom, Lucretius uses it to explain everything, from cosmology, physics, geology, and meteorology, to biology, psychology, sociology, and even political science. The modern reader encounters arguments that seem like impeccable scientific reasoning in some cases and ridiculous speculation in others. The poem requires a certain self-confidence on the part of the reader to resist dismissing the whole on the basis of its fantastic conclusions and to focus instead on its sound foundations and startling intuitive insights.

Lucretius is at his best in illustrating the reasons for believing in atoms in the first place. One of the most compelling is the neat solution to the mind-boggling problem of the divisibility of matter: If you cut a coin in two and then into quarters, then eighths, and continue cutting, where do you stop? If there were no end to this process, at some stage you would come to a piece of metal that was so small that for all intents and purposes it would be indistinguishable from nothingness. But is that piece really still *something*? The mind reels at the notions of the infinite and the infinitesimal, in physics as much as in mathematics. If a drop of milk diffuses in water, does it actually disappear into nothingness? If so, what happens to the bulk of the milk? Does the process of spontaneous subdivision stop somewhere? If so, where?

The concept of the finite "atom," a word meaning "indivisible" from the Greek *a-tomos,* or "un-cuttable," provides natural answers to these conundrums. Long before a piece of matter becomes undetectably small, the process of division stops at the atom. Milk, when it dissolves, simply separates into innumerable molecules that hide in the empty spaces between water molecules. The mind does not need to struggle with the idea of dissolution into infinitesimally small portions, because in nature they don't exist.

Another fundamental argument for atomism adduced by Lucretius concerns the interpenetrability of matter. If matter were continuous, he asks, how could a fish swim through water? If water were a continuous material that fills space seamlessly, how could motion begin? Lucretius explains that believers in the continuity of matter say that

> . . . water gives way to the fish
> As it swims, and opens a passage for it to pass,
> Because there is a space left behind the fish
> Into which the liquid can flow: and this, they say, demonstrates
> How other things can change place, although space is full.

But, he counters,

> This explanation rests on erroneous reasoning,
> For how after all can the fish find a way to move forward
> If the water does not give way to it? And how can the water
> Give way to the fish, unless the fish can move forward?

If you assume that there are atoms and voids, the paradox is resolved: The forward extremity of the fish seeks out the closest gap between neighboring molecules of water and enters this crevice without opposition.

Beginning with atoms, Lucretius proceeds to explain how

they combine to form increasingly complex structures until the entire world is built up. His guiding metaphor is the alphabet: Just as the infinite wealth of literature is produced by the combination and permutation of only twenty-six letters, so the universe, life and all, is composed of countless identical copies of a mere handful of different atoms.

But there is another side to the work—a religious intent. Its stated purpose is to deliver mankind from the oppressive yoke of superstition, and in some of his most passionate and lyrical passages, Lucretius argues that an explanation of natural processes in terms of atoms obviates the necessity of a belief in divine powers. If, for example, the wind is imagined as consisting of a stream of material particles rather than the work of some vengeful and cruel god, then Agamemnon's sacrifice of his daughter, Iphigenia, for the purpose of raising a favorable wind appears in all its barbaric horror as a futile murder.

Unfortunately for his scientific purpose Lucretius succeeded only too well in conveying his moralistic message. During his own lifetime and on into the seventeenth century, he was branded an atheist, and his doctrine was condemned. Atomism was confused with atheism, and the scientific world was thereby deprived of many of its trenchant, and fundamentally correct, explanations. Long before the Church tried to suppress Galileo's science in the name of God, Lucretius wreaked just as much mischief by denying the gods in the name of science. Democritus might have laughed at the endless quarrel between science and religion, which helps neither of them and demeans both. In any case, the seed that he and Leucippus had planted did not perish; in time it revived and grew to become the dominant view of nature that it is today.

The turning point came at around the year 1600. During the age of exploration, as Shakespeare molded our language and the courtly reign of Queen Elizabeth gave way to the more mercantile rule of James I, modern science was born. A momentous innovation had captured the minds of scholars all over the Europe: They had discovered how to discover. People like Johannes Kepler in Prague, Galileo Galilei in Venice, Thomas Harriot in London, and a generation later, René Descartes in Paris, invented a new methodology. They joined two disparate techniques of inquiry into a single discipline that turned out to be vastly more powerful than either one by itself: the rational analysis of Greek philosophy and the empirical investigation of medieval alchemy. From the Greeks they took logic and mathematics; from the medieval adepts, who had gathered insights from astrology, medicine, mining technology, and other practical arts in their desperate efforts to wrench aside nature's veil, the new scientists learned the value of experimentation and quantitative measurement. The combination proved to be fruitful beyond expectation.

Mathematics was the key. The alchemists already knew that understanding nature requires both observation and thought. They also knew how to weigh, measure, combine different quantities of materials, and heat them for different lengths of time to produce different results. The flamboyant alchemist Dr. Philippus Theophrastus Aureolus Bombastus von Hohenheim, called Paracelsus, who was expelled from the University of Basel in 1528 for his temerity in inviting ordinary townspeople to his lectures, one day nailed a syllabus to the door of the university. It contained a promise that flew in the face of medieval practice, and sounds like a recipe for modern science: "If I want to prove anything, I shall not do so by quoting authority, but by experiment and by reasoning thereupon." (How Paracelsus would have loved to see a tame mercury atom! His whole life revolved around that marvelous metal. As physician he had discovered that it cures syphilis, as alchemist he had tried in vain to transmute it into gold, and as charlatan he had occasionally amazed his victims by succeeding.)

But the method advocated by Paracelsus was not yet physics, not until Kepler and the others learned to cast the results of observations in quantitative form and to correlate them with each other by means of mathematics. Only then, when Galileo found a simple formula for the distance a cannonball falls in a given time, Kepler an equation for the orbit of Mars, Harriot a mathematical rule for the bending of light in water, and Descartes a numerical derivation of the size of the rainbow, did modern science begin.

Atomism was there, at the birth of the new science, but it still suffered from the old stigma of its association with atheism. Thomas Harriot, a pivotal figure both historically and geographically, whose life straddled the year 1600 and who spent a year in America as a member of the first English colony in Virginia fifteen years before the settlement of Jamestown, was a confirmed atomist. In spite of his many great discoveries in physics, astronomy, and mathematics, his name is virtually unknown. He published very little, chiefly because he was formally, albeit unsuccessfully, charged with atheism, a serious felony at the time, and didn't dare to expose himself—a late victim of Lucretius's zeal.

Thomas Harriot knew what had to be done with atoms, but he did not have the requisite background to do it. On the second of October 1606 Kepler wrote to him to seek advice on the theory of the rainbow. Two months later Harriot replied with a description of the bending of light in a material medium: Reflections of light from the atoms of the medium alternate with its unimpeded travel through the interstitial vacua. Harriot's final advice, astonishing for its depth of scientific and human insight, could serve as a motto for all who seek an answer to Thales' question: "I have now conducted you to the doors of nature's house, where its mysteries lie hidden. If you cannot enter, because the doors are too narrow, then abstract and contract yourself mathematically to an atom, and you will easily enter, and when you have come out again, tell me what miraculous things you saw."

What Harriot had proposed was a thought experiment, a theoretical exercise the Germans call *Gedankenexperiment.* For Kepler, who had one foot still firmly planted in medieval philosophy, the notion of climbing inside a hypothetical atom was too radical, and he rejected Harriot's suggestion. He shouldn't have, as it turned out, because the thought experiment was to become one of the most powerful techniques in the tool chest of twentieth-century atomic physicists.

After Harriot's death in 1621, while modern physics was

gearing up to its first supreme triumph—the Newtonian theory of mechanics—atomism finally lost its association with atheism. About forty years later, when young Isaac Newton began to study physics, he could write without fear of reprimand that "the first matter must be attomes." He believed in atoms and based much of his philosophy on the assumption that they exist, but he, too, lacked direct evidence to support this hypothesis. He prepared the ground for those who were to follow, however, by perfecting the laws of motion of material particles: how they bounce off walls and off each other, how they are affected by external forces, how they change direction and speed, in short, how they behave. When David Wineland designed his atomic trap three hundred years later, he used Newton's laws in their original form, not relativity or quantum theory or any of those modern refinements, to compute the motion of each mercury atom. Newton would have understood perfectly how atoms are trapped, although he would have been as astonished as I was to see them blink.

The terms in which the mature Newton expressed his belief in the atomic hypothesis are clear and strangely prescient:

> Now, the smallest particles of matter may cohere by the strongest attractions, and compose bigger particles of weaker virtue; and many of these may cohere and compose bigger particles whose virtue is still weaker, and so on for divers successions, until the progression ends in the biggest particles on which the operations in chemistry, and the colors of natural bodies depend, and which by cohering compose bodies of a sensible magnitude.

This passage astonishes, not only by its wealth of details about the properties of atoms but also because it turns the hierarchy of matter established by Democritus and Lucretius on its head. The building blocks of chemistry and of optics, which is to say the atoms of the ancient philosophers, are not the *smallest* particles in nature, but the *largest.* Newton guessed correctly that they are in turn made of more elementary components, and those of yet smaller ones. Furthermore he sensed that the forces between particles should increase in strength with each step down the ladder of complexity. Today the most elementary particles at the bottom, the truly primitive indivisibles, are called quarks and leptons, and the forces among them are indeed much more powerful than the forces between atoms. Newton did not predict quarks, but his instincts about the architecture of matter were sound.

Eleven years after Newton's death, in 1738, his laws of motion were finally brought to bear on atoms, invisible though they were, by Daniel Bernoulli of Basel, in order to make a quantitative prediction of an observable phenomenon. It had taken a long time for mathematics, the key to modern science, to catch up with the atomic hypothesis. By postulating that a gas consists of innumerable solid particles that are far apart from each other, rarely collide, and dart about in straight lines except when they bounce off walls like tennis balls, Bernoulli was able to explain its basic properties. In particular he was able to show that the pressure of a gas must vary inversely with its volume, which is precisely what is found experimentally.

Pressure, in this model, is nothing but the push of countless atoms hitting and recoiling from each wall. Bernoulli considered a cubical container filled with gas and noticed that if the length of each side of the box were doubled, it would take an atom twice as long to travel across the box and back again, so that the number of collisions of atoms with one wall, during each second, is cut by half. At the same time the wall has become four times larger, so that the force per square inch, or pressure, is reduced by another factor of four. Thus the pressure decreases by a factor of eight while the volume increases by the same ratio.

The significance of this feat of mathematical analysis can hardly be overstated. Unlike Democritus, who introduced atoms as convenient logical devices, Bernoulli thought of them as actual bits of matter that make their presence felt by acting collectively in enormous numbers. Certain observable effects, such as the volume-pressure relationship in gases, do not depend on the sizes and structures of the atoms, while other phenomena, such as the laws of optics, do. It was Bernoulli's genius to pick a case in which a quantitative prediction followed mathematically from very general assumptions. In this respect his simple and straightforward derivation is the first example of modern theoretical atomic physics.

Nevertheless the line of investigation opened by Bernoulli was not sufficient to establish the reality of atoms. It was still possible, then and for more than a century afterward, to think of atoms the way the church preferred to think of the heliocentric theory: a neat trick to furnish the right answers, but not the real thing. ("Galileo will act prudently," wrote the kind-hearted Cardinal Bellarmino, ". . . if he will speak hypothetically. . . . To say that we give a better account of the appearances by supposing the earth to be moving, and the sun at rest . . . is to speak properly; there is no danger in that, and it is all that the mathematician requires.") Nature, in this evasive manner of speaking, behaves *as if* the earth moves, and as if gas consists of atoms, but that's not necessarily the way things really are. The "as if" escape clause always lurks in the background of any description of nature. In our own time it is in full force to help physicists cope with the paradoxes of quantum physics. In the eighteenth century, Daniel Bernoulli's theory notwithstanding, it was routinely invoked to deny the reality of atoms.

Real or not, atoms are physical phenomena, and physics in the Age of Reason required that they be examined in accordance with the Newtonian philosophy. The strength of Newton's method lay in its two-pronged approach to the description of nature. On the one hand, general theoretical assumptions lead, by means of deduction, to particular predictions that can be verified experimentally. Bernoulli's derivation of the volume-pressure relationship from the atomic hypothesis is a classic example of deductive reasoning. Induction, on the other hand, proceeds in the opposite direction, from particular empirical evidence to the general law of nature. Newton's method was crowned with success when the two paths converged on the same results.

The inductive approach to the study of atoms was pursued in the late eighteenth and early nineteenth centuries by the chemists. Where Bernoulli had focused on the motion of atoms, with no reference to their weights and modes of interaction, the chemists attempted to do the opposite; they put the emphasis on combinations and separations, with no regard to motion, and thus developed their own theory of the chemical atom, with scant relationship to Bernoulli's conception of the physical atom. At that time the deductive and inductive methods applied to atoms proceeded simultaneously and showed little tendency to converge. Like hunters in tall grass stalking the same prey from different directions, physicists and chemists attacked the problem of atomic structure in their separate ways.

The chemists' conception of the atom began in the early nineteenth century with a system of shorthand symbols for the elements. The purpose of the scheme, at first, was to facilitate the bookkeeping required for recording the growing body of chemical knowledge. The most influential proponent of this system, the English schoolteacher John Dalton, preferred pictograms resembling the astronomical signs for planets, but for typographical reasons these were soon replaced by the familiar letters H for hydrogen, O for oxygen, Au for gold, and so on. The important discovery that water consists of two parts hydrogen and one part oxygen (the word *part* being suitably defined) is neatly summarized in the formula H_2O, but nothing in this notation refers to physical atoms.

Although Dalton himself believed in the reality of atoms when he began to develop an atomic theory of chemistry in 1803, many of his contemporaries did not. Like others who came before and after him, they regarded atoms as a convenient fiction and thought of them the way we think of shares traded on the New York Stock Exchange, as useful abstractions rather than tangible objects. The realization that the letters *H* and *O* stand for real, physical bits of matter called atoms came slowly.

It may seem strange to us today that the successes of the philosophical atom of Democritus, the physical atom of Bernoulli, and the chemical atom of Dalton were not sufficient to establish the reality of atoms. Recent history provides a parallel that sheds light on the frame of mind of the scientists of the previous century. When the American physicists Murray Gell-Mann and George Zweig proposed the idea of quarks in 1964, they were thinking of mathematical objects which would enable them to make sense of the welter of experimental information about the newly discovered subatomic particles that was becoming available at the time. These quarks resembled Dalton's chemical atoms in that they were the primitive units which, by combination and permutation, made classification possible. They were, in the image of Lucretius, the alphabet of nature.

Later in the same decade an altogether different kind of particle was proposed. The American physicist Richard Feynman suggested that atomic nuclei might contain real, hard kernels of matter, like apple seeds. Since they were parts of other particles, Feynman somewhat inelegantly dubbed them partons, and for a while partons and quarks coexisted as separate theoretical conceptions. In the end, however, as confidence grew that quarks were more than metaphorical aids to understanding, and as evidence for the existence of partons was actually found at the two-mile linear accelerator in Stanford, the two concepts finally coalesced into one: Partons *were* quarks, and the term *parton* gradually disappeared from the vocabulary as a separate concept. The early quark hunters stalking their prey from different directions were in the same predicament as the atomists of the nineteenth century.

By the year 1900 indirect evidence for the existence of atoms, derived from chemistry and the theory of gases, as well as from other developments, such as the emerging theories of heat and electricity, had become overwhelming. Nevertheless some leading physicists continued to use the "as if" defense even then, because they refused to accept what their senses did not reveal to them directly. The famous Austrian physicist Ernst Mach, whose name is used as a measure of supersonic speed, and whose philosophy of science deeply influenced Einstein, held out longer than most. "Atoms cannot be perceived by the senses . . . they are things of thought," he wrote, and "molecules are merely a valueless image." More charitably, he allowed that "The atomic theory plays a part in physics similar to that of certain auxiliary concepts in mathematics; it is a mathematical model for facilitating the mental reproduction of facts."

Then, in 1903, a simple instrument shaped like the eyepiece of a telescope and called a spinthariscope, from the Greek word for spark, was invented. Inside it a fluorescent surface, like a tiny version of a video monitor, was irradiated by a stream of particles from a radioactive source. Each collision of one of these particles with the screen was marked by a flash of light, a twinkle that announced the decay of an individual atom in the source. Mach's assistants dragged the sixty-five-year-old doubting Thomas down to the laboratory to witness this marvel, which to them represented visible evidence for the reality of atoms. Mach squinted into the apparatus, saw the scintillations, and surrendered. "Now I believe in the existence of the atom," he said, although six years after his apparent conversion, the old skeptic reverted to his deeply held prejudice and again referred to atomism as "hypothetical-fictive physics."

By this time, however, Mach's opinion no longer made any difference: around 1900 the atom, which had been conceived twenty-three centuries earlier on the sunny shores of the Aegean, finally left the realm of hypothesis and was accepted as fact. In science the transformation of a hypothesis into fact is not an instantaneous event. An idea usually gathers experimental support, and therewith acceptance by the scientific community, over an extended period of time. A plot of "acceptance," which increases from zero to a hundred percent on the vertical axis, with years along the horizontal, would be a graph that increases steeply at first and then flattens out as it approaches certainty, like the profile of a mesa viewed from a distance.

Consider, for instance, Newton's theory of gravity, the celebrated law of universal gravitation, which describes the mutual attraction of all massive bodies to each other. After its formulation in 1666 its acceptance rose quickly as it recorded successes in explaining the complex motion of the moon, the description of planetary orbits, the cycle of the tides, the gradual motion of the celestial north pole away from the North Star, and the dance of the moons of Jupiter. By the end of the century acceptance had reached, say, 90 percent. But there were still unbelievers, and alternative theories continued to be defended.

When, in the middle of the eighteenth century, the theory successfully predicted the return of Halley's comet, its acceptance rose a few more percentage points. Then, a hundred years later, the planet Uranus appeared to be straying from its computed orbit, and doubts about the universal validity of the law surfaced again. But then the young English astronomer John Adams, and independently his French colleague Joseph Leverrier, used that very law to predict the existence of a hitherto unsuspected planet, which they blamed for Uranus's aberrations. When the new planet Neptune was discovered in 1846, exactly where they said it was supposed to be, acceptance became certainty: the law of gravitation became a fact.

The atomic hypothesis took much longer to travel the same road. The remoteness of atoms from ordinary sense experience prevented it from reaching full acceptance for millennia. The contributions of Leucippus and Democritus, of Lucretius, of Harriot, Newton, and Bernoulli, of Dalton, and of other renegades who could envision what their eyes could not see and their hands could not feel, who could somehow divine the innermost secrets of nature, spanned the entire history of Western civilization. The taming of the atom that had begun with the laughing philosopher in 400 B.C. ended with the spinthariscope of 1903. But at the very time that atomism was finally accepted as an established fact, the atom itself began to fly apart.

Without pausing to savor the successful end of the ancient quest for understanding the nature of matter, physicists immediately pushed on to the next problem, the central question of twentieth-century atomic physics: What is the structure of the atom? or, more precisely, What are the constituents of the atom, and how do they fit together?

2

The Components of the Atom

The two principal building blocks of the atom, the electron and the nucleus, were discovered in England at the turn of the century, in 1897 and 1910 respectively, the first at Cambridge and the second at Manchester. Although the scientists who made these discoveries, Sir Joseph John Thomson and Lord Ernest Rutherford, were lifelong colleagues and friends, their contributions provide a study in contrasts. The discovery of the electron in the waning years of the nineteenth century represented the culmination of the methodical electrical researches that began in the eighteenth, whereas the nucleus was found by accident in the course of the investigation of radioactivity, a new and unexplored phenomenon at the time. It was fitting that the dates of these two milestones straddled the year 1900: As the world crossed from one century into the next, physicists finally managed to cross the surface of the atom and slip inside.

The object in the middle, the nucleus, is actually not a fundamental particle at all but a composite object that differs from atom to atom. It turns out, however, that it is so small that its inner structure doesn't affect an atom's architecture. Only the gross, external properties of the nucleus matter—its total mass, its electrical charge, and its magnetic strength. From the point of view of the electrons, the nucleus looks like Earth seen by Voyager 1, or like a mercury atom in a trap illuminated by ultraviolet light: It is a minuscule corpuscle, a miniature grain of sand in the center of the atom.

The electron, on the other hand, is fundamental. Unlike the nucleus it has neither shape nor structure nor constituent parts. It also differs from the nucleus in that its discovery has a prehistory: it is, in fact, the long-sought-after particle of electricity. As early as 1750, Benjamin Franklin, in the tradition of atomism, had surmised that an electrical current consists of a stream of extremely subtle particles that are, alas, too small to be perceived. They owe their actual discovery a hundred and fifty years later to a simple technological innovation that also dates to the time of Franklin and has transformed the world as much as the legendary invention of the wheel.

Throughout the eighteenth century the two principal scientific instruments were the electrical machine and the air pump. The former produced sparks by rubbing leather on glass, and the latter a vacuum by sucking air out of a sealed vessel, the way a hypodermic syringe sucks blood out of a vein. The pivotal innovation came about when these two simple devices were combined for the purpose of showing "the appearance of the spark in vacuo," in the jargon of the time.

A spark in air, which is a bolt of lightning in miniature, is an unruly affair because the electrical current must force its way through a mass of air molecules, like a police patrol wading through a mob. Remove the air, however, and the stream of electrical particles becomes smooth, swift, and tame. Whereas electricity blazing across a stormy sky, or coursing through a metallic wire, represents raw power, an electrical current in a vacuum exemplifies the exercise of delicate control.

The marriage of the electrical machine to the air pump produced an impressive list of discoveries and inventions: neon tubes and fluorescent light bulbs; the vacuum tubes used in early radio, TV, and computers; cathode-ray oscilloscopes, TV picture tubes, and computer monitoring screens. In short, almost every facet of our information-sodden society is in some way dependent on electricity flowing through evacuated tubes. More fundamentally, X-rays were discovered when an electrical current in a vacuum tube was forced to crash into a block of copper, and modern particle accelerators are elaborations of the same technique. The two-mile linear accelerator at Stanford University, which was instrumental in the discovery of quarks, is, in principle, nothing more than an enormous, scaled-up version of the little vacuum tubes of the eighteenth century. But the most significant legacy of that homely apparatus was the discovery of the electron by Joseph John Thomson.

J.J., as everyone including his son called him, was an unusual experimentalist. Unlike most practitioners of his calling, he was so clumsy with his hands that his assistants discouraged him from actually touching any of the equipment in his laboratory. His student F. W. Aston, later a famous physicist in his own right, recalled the ill-shaven, wizened little professor in the rumpled suit this way:

> When hitches occurred . . . along would shuffle this remarkable being who, after cogitating in a characteristic attitude over his funny old desk in the corner and jotting down a few figures and formulae in his tidy handwriting on the back of somebody's Fellowship Thesis or an old envelope or even the laboratory checkbook, would produce a luminous suggestion like a rabbit out of a hat, not only revealing the cause of the trouble, but also the means of the cure.

Though lacking in dexterity, J.J. clearly had an intuitive feeling for experimental apparatus, and a genius for drawing profound conclusions from simple observations.

At the turn of the century it would not have been possible to isolate an electron: the technology for trapping single particles was not perfected until eighty years later. What J.J. did instead was a clever variation on the "spark in vacuo" experiment. He directed a stream of electricity, about as thick and long as a knitting needle, through an evacuated glass tube at a fluorescent screen of the kind used six years later in the spinthariscope that had demonstrated the reality of atoms to Ernst Mach. Where the beam struck the screen, it produced a small pinpoint of light. By placing a simple little magnet on one side of the tube, he was able to deflect the stream of electricity so that the dot jumped downward.

From the displacement of that shimmering spot by a fraction of an inch J.J. drew a conclusion that opened a door on the world of subatomic physics. He explained his observation by assuming that electricity consists of tiny electrically charged corpuscles called electrons. By comparing their common trajectory through his apparatus to the path of a cannonball, with the magnetic force replacing the force of gravity, he was able to calculate the deflection of the dot from a knowledge of the strength of the magnet. In the course of the calculation J.J. had to assume a value for the mass of the electron. By varying the value he assigned to it until his calculation matched the result of his experiment, he found it to be about two thousand times smaller than the mass of a hydrogen atom, which chemists had determined several years earlier. (The actual calculation was a bit more convoluted, but the result was the same.)

The intellectual impact of Thomson's discovery was stunning. For centuries atoms were thought to be the most basic constituents of matter, and hydrogen was the lightest among them. Yet here was something two thousand times lighter. The deconstruction of the atom into subatomic particles had begun.

Although Democritus had been wrong about the details—there *is* something more basic and fundamental than the atom—he was right in principle. What he could not have foreseen was that when the search for the indivisibles of nature reached the atom, it didn't stop there but continued on to the electron and the nucleus, and then to the constituents of the nucleus, and eventually down to quarks. Ironically it turned out that the first subatomic particle to be discovered was also one of the most fundamental. Of the hundreds of so-called elementary particles, the electron is one of the very few that has remained impervious to further subdivision. It is truly elementary.

With the discovery of the electron J.J. ushered in twentieth-century physics, but then the pace of discoveries quickened and he dropped back. He opened the door into the atom but never went through himself, while his assistant and collaborator Ernest Rutherford charged through it like a bull. The two men differed not only in their approach to physics but also in everything else, even physical appearance. In one photograph they look a little bit like Laurel and Hardy: The slight, stooped, bedraggled J.J., his coat buttoned to the top, his bowler hat pulled down to the eyebrows, listening tightlipped while Rutherford, huge, handsome, relaxed, and unbuttoned, looms over him, talking with his usual intensity.

In 1897, the year in which the electron was discovered, Rutherford was working in J.J.'s laboratory as a junior assistant. But instead of following his mentor in the study of the electron, the young Rutherford turned his attention to the science of radioactivity, which had just been discovered in France. His decision to break away from his teacher's conservative course in order to strike out into new territory was the brash act of a youthful and self-confident scientist. Later he was to call that move the most important of his life, for eventually it led to his greatest discovery, the atomic nucleus.

Rutherford's adventurousness also took him beyond his mentor in another way: Where J.J.'s strength had been the ability to visualize the interior of his experimental apparatus, Rutherford was able to follow Thomas Harriot's advice and contract himself down to atomic size and look around. When someone suggested that atoms might not exist, he insisted, in his uncommonly loud voice, that they did, that they were "jolly little beggars, so real that I can almost see them." If he could only have known that his imagination put in motion a course of developments that made atoms visible three generations later.

In 1909 Rutherford, then at the University of Manchester, was studying the manner in which alpha particles, a form of positively charged radioactive rays, scatter when fired at diaphanous gold foils. Alpha particles, which emanate naturally from radioactive materials such as radium, are so energetic that they normally shoot right through thin targets. Most of them continue on their original straight path; only a few manage to glance off gold atoms and are thereby deflected to the side. According to Rutherford, an assistant came to him one day and asked, "Don't you think that young Marsden whom I am training in radioactive methods ought to begin a small research?" "Now, I had thought so too," Rutherford continued, "so I said, 'Why not let him see if any alpha particles can be scattered through a large angle?' "

Ernest Marsden tried the experiment. He placed a piece of radioactive material inside a heavy leaden casket which was pierced by a small hole that allowed alpha particles to escape in a thin, straight beam. These particles hit a piece of gold foil placed in their path, and on the other side of this target Marsden had placed a spinthariscope: each scintillation on its fluorescent screen signaled the arrival of a single alpha particle. In this manner he found that most of the alpha particles arrived on the far side of the gold foil practically undeflected. Then he brought his detector around to the near side, the side from which the beam entered the gold foil; he was looking for reflections, and expected to find none. But although the number of particles he counted there was indeed small, a few did recoil from the gold and reverse course.

To Rutherford, who had a firm grasp of the sizes, speeds, and masses involved in the atomic realm, this result was astonishing. He said later that "It was as though you had fired a fifteen-inch shell at a piece of tissue paper and it had bounced back and hit you."

He deliberated over this perplexing result, and a year later, just before Christmas 1910, the answer came to him. By imagining what an encounter between an individual alpha particle and a gold atom would look like, he concluded that the agents responsible for this phenomenon must be very tiny, heavy, positively charged nuggets hidden within each gold atom, and that these repelled the projectiles the way shrapnel buried in a wooden board would scatter steel bullets. Rutherford imagined that every atom contained one of these nuggets and called it the nucleus.

To arrive at this conclusion Rutherford calculated the path of a hypothetical alpha particle through the vacuum between gold atoms, into the interior of one particular gold atom, and back out again. As the positively charged alpha particle approaches the positively charged gold nucleus, it is repelled more and more strongly, because like charges repel each other. Like a fugitive's car approaching a police barrier, it makes a U-turn and quickly leaves the atom altogether. Rutherford's theory resembled J.J.'s calculation of an electron's curved trajectory through a vacuum tube, only the arena had shifted from the laboratory, where the events were still visible, down to the inaccessible realm of the atom. Henceforth that invisible world would be the physicists' principal domain.

The revolutionary implication of the discovery of the nucleus is hard to overestimate. Five years earlier Albert Einstein had published the special theory of relativity, thereby destroying the rigid framework of space and time that had been the skeleton of classical physics. But while Einstein had merely demolished an abstract idea, Rutherford had dissolved matter itself; the atom that was imagined as a tangible object, as a miniature grain of sand, could no longer exist. He had transformed it into a ball of vacuum in which a minuscule nucleus and a few pointlike electrons dance an arcane minuet. Rutherford pulled the rug, as well as the rigid floor and the solid ground beneath it, out from under our feet and left us walking on a void. Thus he contributed even more than Einstein to the overthrow of the coherent worldview that had dominated physics virtually unchanged since classical antiquity.

With the discovery of the nucleus the primary constituents of the atom were in place. Every element in the periodic table is characterized by a nucleus with a different positive charge, and each nucleus is surrounded by a retinue of negative electrons whose combined charge precisely balances it. Thus the hydrogen atom has one positive charge in the center and one electron; the helium atom has two of each; and the lithium atom, which is portrayed on the ten-drachma coin, has three. The electrons are held in place by electrical attraction and contained within an electronic sphere that is ten thousand times larger than the nucleus, but two thousand times lighter. This much is clear, simple, and indisputable. The question is how, exactly, are the electrons arranged around the nucleus? This is the problem of modern atomic theory.

Since atoms are too small to be dissected like frogs, the problem can only be attacked by indirect means. Rutherford had opened one door into the atom, but his method was too brutal: On their way in, alpha particles carried so much energy that the delicate fabric of the electronic veil, with a million times less energy to hold it together, was torn to shreds like tissue paper by a cannonball. Light rays are a better substitute for alpha particles because they are far less destructive, but first it would be necessary to understand more about light itself.

Newton, in the proper tradition of atomism, believed that light consists of particles but could do no more with that hypothesis than he could with the atomic theory of matter itself. A hundred years later light was found to behave more like waves than individual particles, and by the end of the nineteenth century it was generally agreed that the particle theory of light was untenable.

The principal evidence for the wave nature of light is the phenomenon of destructive interference. Two crossing streams of water from garden hoses, or two bursts of bullets from machine guns cannot annihilate each other; they can only enhance each other's effect. But waves are different; if you throw two pebbles into a pond and watch the ripples, you will find places on the water where the waves from the two sources cancel each other out, where a crest from one meets a trough from the other, and vice versa. This pretty effect is called destructive interference and can be summarized by saying that under certain circumstances wave plus wave equals nothing. Particles can't do that.

A convincing example of the destructive interference of light waves is no farther away than the tip of your nose. By squinting into the distance, or at a light bulb, between the knuckles of your extended fingers when they are held an inch in front of your eye and almost touching each other, you can see vertical black striations. These are not caused by shadows, nor are they images of your eyelashes; they are genuine optical phenomena and explainable only in terms of light waves. In some places portions of light that stream through the gap happen to be out of sync with other portions that arrive in the same place, and the two cancel each other out. When experimental evidence of this kind was further supported by James Clerk Maxwell's theory of 1873, which described light waves as consisting of oscillating magnetic and electrical (or, more succinctly, electromagnetic) fields, and which accounted for all the subtle optical phenomena known at the time, the issue seemed settled in favor of waves.

Interference experiments carried out with different sources of light even led to a quantitative, numerical measure of color, the most appealing quality of light. Color is related to wavelength, the distance between successive crests of a wave: The longer the wavelength, the redder the light; the shorter the wavelength, the closer it is to the violet end of the rainbow. An alternative way of treating color numerically is by means of frequency, or the number of crests of a wave that pass an observer per second. Red light, with its long wavelength, has a low frequency; blue and violet light are characterized by short wavelengths and high frequencies. Thus, the color of light came to be associated with wavelength and frequency, which are two closely related quantitative features of waves.

But atomism again proved its persistence and returned to reclaim even light. The first step toward a particle theory of light was taken in the first year of this century by the German physicist Max Planck. Planck, who is regarded as the grandfather of the quantum theory, was a reluctant revolutionary. At age forty-two he was professor of physics in Berlin and had no intention of doing anything but furthering the classical program of physics—explaining the phenomena of the material world in terms of the mechanical and electromagnetic theories of Newton and Maxwell. Fate had something else in store for him.

The problem Planck was struggling with at the time did not seem very profound. He was trying to explain how a hot object changes color with temperature—how the element of an electric range glowing in the dark changes from black to red to white-hot. His formula for associating temperature with color did not work at the violet, high-frequency end of the spectrum. Classical theory predicted that short wavelengths should be more plentiful than long ones: If you count waves on the ocean, you find many tiny ripples for every long roller. But in the experiments that Planck was trying to understand, the high frequencies of a hot body didn't show up as expected—some mysterious mechanism was suppressing them. Casting around for a way out, Planck thought that if, for some unknown reason, it takes more *energy* to generate high frequencies than low, then those violet colors won't appear, simply because there isn't enough energy available to produce them.

The idea was, as he called it later, an "act of desperation" that had no physical basis. For the rest of his life Planck would try in vain to fit it into the framework of classical physics. But it worked. Not only did it produce a formula for the relation between color and temperature that was in perfect agreement with the evidence, but as the cornerstone of modern atomic theory it earned Planck the Nobel Prize in 1918.

In order to express his idea coherently, Planck had to make the perplexing assumption that a hot body emits light and radiant heat not in a continuous stream, as common sense and classical physics had led him to expect, but in discrete bundles of energy. Each bundle carries an amount of energy that is a multiple of its frequency—the higher the frequency, the higher the energy. The formula for calculating the energy of a bundle of light from its frequency is called Planck's law. The constant that accomplishes the conversion is Planck's constant, today the very linchpin of atomic physics.

Twenty years ago the German government honored Planck's achievement by placing his portrait on the two-mark coin, where he appears in profile above his vital dates 1858–1947. With an aquiline nose above a drooping mustache he stares balefully through tiny wire glasses at a world he had helped to transform but had never quite succeeded in accepting. His bald skull seems huge, as though distended with sheer intelligence. Unlike the ten-drachma coin, which bears the portrait of Democritus and an atomic icon, the two-mark piece gives no indication of Planck's contribution to science; there is no obvious way to convey his discovery in visual terms. But in spirit, the accomplishments of the two men were complementary: They were the seeds of the momentous idea that the world of matter and light, unbroken and continuous though it appears, is really divided into discrete units, like infinitesimal coins.

In 1905, the same year in which he also invented the theory of relativity, Albert Einstein pointed out the revolutionary implications of Planck's discovery. He was interested in the way light shining on certain metals can release electrons, an effect that underlies the operation of electric eyes. What puzzled him was that even an extremely feeble ray of light releases surprisingly energetic electrons. The process seemed as paradoxical as if a gentle ocean wave (the light) were suddenly to hurl a piece of flotsam drifting on its surface (the electron) high into the air. Einstein proposed that the image is wrong, that light does not resemble gentle waves on an ocean so much as a volley of cannonballs and that its interaction with electrons should be understood in terms of the effect of those powerful projectiles on the flotsam. Specifically he suggested that light comes in the form of bundles, or corpuscles, which he called light quanta. In his view feeble light consisted of a sparse beam of quanta, each of which packed enough energy to knock an electron out of a metal.

Einstein's idea built on Planck's theory, but differed in one crucial respect. Where Planck had been forced to assume that light is *emitted* by hot bodies in the form of discrete bundles, Einstein maintained that it always *consists* of such bundles. He graphically illustrated the difference by comparing light with beer: Where Planck had merely shown that beer is customarily *sold* in pint bottles, Einstein suggested that it actually *consists* of indivisible pint portions—a far more radical proposition. The amount of energy carried by each quantum is related to its frequency by Planck's law. Henceforth color, wavelength, frequency, and energy became almost interchangeable concepts for describing light.

The wave theory of light was so powerfully held at the time that Einstein displayed an uncharacteristic diffidence in reintroducing an atomistic interpretation. Falling back on the old "as if" dodge, he called his theory heuristic, meaning provisional, or helpful as an aid for further investigation. By a curious coincidence Isaac Newton, in a 1672 address to the Royal Society in which he explained his belief that light consists of particles, also labeled the idea heuristic. But even with that qualification, the scientific community reacted with skepticism to Einstein's theory. In 1913, eight years after the publication of his theory of light quanta, Planck and three prominent colleagues wrote a recommendation for the appointment of Einstein to the Prussian Academy of Sciences that included the remark, "That [Einstein] may have occasionally missed the mark in his speculations, as, for example, with his hypothesis of light quanta, ought not to be held too much against him, for it is impossible to introduce new ideas, even in the exact sciences, without taking risk."

But, as in so many other instances, Einstein's idea proved to be fruitful, and survived. His light quanta were later called photons (from the Greek roots for "light" and "thing") by the American chemist Gilbert Newton Lewis, and the name stuck. The true nature of the photon—whether it is a particle, characterized by its energy, or a wave with a definite frequency, or both, as implied by Planck's relationship between frequency and energy—is not much of an issue for professional physicists, because they have a complete theory for the interaction of light with matter in which all energy comes in discrete, irreducible bundles called quanta. When the energy happens to be carried by light, a quantum is called a photon. The scintillations I saw on David Wineland's monitor, even before I spotted my first atom, signaled the arrival of individual photons of ultraviolet light on the screen, and provided visible proof that Einstein's speculation had been on the mark.

Electron, nucleus, photon—these are the three coins of the realm of atoms. Atoms shake off photons and thereby reveal their structure. When an atom is struck by a particle such as an electron, a photon, a nucleus, or even another atom, it sometimes responds by emitting photons. Each chemical element is capable of generating only photons of a few specific frequencies, or colors. For the chemist these discrete colors, collectively known as the atomic spectrum, constitute a unique fingerprint and allow the optical identification of substances. For the physicist the spectrum is the doorway to an atom's interior.

The difference between an atomic model that accounts for the spectrum and one that doesn't is like the difference between a piano and a garbage can: Strike the former, and you get definite, clear, individual frequencies of sound; strike the latter and you get noise. Just as the musical tones that issue from a piano are clues to its construction, the spectrum of an atom reveals something about its interior design. But by the same token, it is far more difficult to build a piano than a garbage can.

The first successful attempt to describe an atom that would emit a spectrum of light in accord with experiment was made in 1913 by the young Danish physicist Niels Bohr. After obtaining his degree in Copenhagen, he had come to Cambridge to work with J. J. Thomson, but the relationship between the two men didn't jell. For one thing Bohr was not particularly interested in experimentation. More fundamentally he felt that J.J., though "a genius who showed the way for everybody," was the wrong kind of genius. His strength lay mainly in the flair with which he overcame difficulties; he didn't share Bohr's passion for uncovering fundamental truths. So Bohr left Cambridge after a few months and went to work in Manchester with Rutherford, who was then refining his theoretical analysis of Marsden's experiment in terms of the nuclear atom.

Professor Rutherford and his theoretical assistant were well matched. Rutherford was a big, mature man, but in spirit he always remained boisterous and youthful, whereas Bohr was just the other way around. With his full mouth and pendulous cheeks he looked like a precocious cherub, but his mind was that of a venerable philosopher. Rutherford encouraged Bohr in his effort to construct a model of the atom that could be reconciled with the dictates of classical mechanics on the one hand and Planck's law on the other. Although the solution Bohr proposed was too radical to be entirely plausible to the older man, scientific differences didn't prevent the two scientists from developing a warm, lifelong friendship.

Even though Bohr's planetary model has long since ceased to be scientifically relevant, its early successes were so impressive that it continues as the dominant popular image of the atom, as its instant recognizability on the ten-drachma coin demonstrates. In the Bohr model, electrons are caught in the grip of electrical attraction, and circle the nucleus the way planets circle the sun. But the resemblance between the two systems ends there. The solar system is not quantized: the distances of the planets from the sun are accidents of the manner in which the solar system came into being, and it would be possible, for example, to put an artificial satellite into an orbit between the earth and Mars, or anywhere else. In Bohr's model this is impossible. Only a few special, discrete orbits are allowed; all others are forbidden.

In order to assure quantization, which is to say to guarantee that certain well-defined orbits were allowed and all others forbidden, Bohr had to go beyond the model of the solar system. He had to add a new restriction to classical mechanics, a novel limitation that only operates in the atomic realm. He found his new rule by assuming that Planck's law, which relates the energy of a photon to its frequency of oscillation, can also be applied to the energy and frequency of rotation of an electron in orbit—an extrapolation far beyond Planck's original proposal. Bohr used this connection to calculate the allowed orbits of electrons and called it his quantization condition. The formula was a kind of mathematical template for designing orbits, and, like Planck's law, had no theoretical foundation in Newtonian mechanics. Its only virtue was that it worked.

In Bohr's planetary model of the atom each electron travels around the center in a specific orbit with a definite and easily computed speed and energy. Thus the total energy contained in an atom must have a specific value: Its energy is quantized into discrete levels. A stone resting on a staircase is similarly constrained to carry a well-defined level of stored energy (in contrast to one placed on a ramp, which can assume all intermediate levels from top to bottom). With each step through which the stone is lifted it acquires more stored energy, which it would release if it were to fall back to the ground. The same is true for an electron: The larger its orbit, the higher its energy level. When an electron jumps from one orbit to a larger one, it must absorb the appropriate increase in its energy; conversely, when it jumps into a lower orbit, it releases energy.

Usually such jumps are caused by light. A photon delivers a quantity of energy when it is absorbed, or carries it off when it is radiated. The staircase structure of the atom guarantees that only light with discrete energies can be emitted or absorbed by an atom, so that only certain discrete colors are emitted or absorbed. In other words, the atom has a unique spectrum. When the spectrum of hydrogen Bohr had calculated with the help of his quantization condition and Planck's constant agreed exactly with the observed spectrum, the riddle of atomic structure seemed to be solved.

Bohr's model marked another triumph for atomism, which holds that the structure of every physical phenomenon can be understood in terms of the motion of particles. The fact that the method of atomism is thus applied to the atom itself causes a mild semantic confusion: The word *atom* as used by modern scientists does not stand for the Democritan final, indivisible particles, but for the building blocks of matter, which are themselves composite. A couple of centuries earlier Isaac Newton had already anticipated the difference between ultimate indivisibles and the objects that "compose bodies of a sensible magnitude," that is, our atoms and molecules.

According to Bohr's model, atoms consist of electrons and nuclei in mutual electrical embrace in a vacuum. Photons, in this scheme, are not actually constituents of matter but simply the messengers that carry energy into and out of the atom's interior. Of course Bohr's theory still left many loose ends—the obscure quantization condition, the dual nature of the photon as wave and particle, and the mystery of the quantum jumps that had caught me by surprise when I saw them in Colorado—but these were details. The architecture of the atom appeared at last to have come into view.

Much to his disappointment the cogent picture of the material world conjured up by Bohr began to unravel almost as soon as it was proposed. Even as physicists throughout the world labored, with very scant success, to extend Bohr's theory to the elements beyond the simple hydrogen atom, as well as to other microscopic phenomena, a young visionary was nursing bigger dreams. In 1924 Prince Louis de Broglie (rhymes with *Troy*), the reclusive younger son of a French duke who had come to physics late in life, and then only because his brother already had a successful career in it, submitted a thesis to the University of Paris in which he made a radical proposal.

De Broglie pondered the relationship between frequency and energy expressed by Planck's law. For the third time in a quarter of a century this simple little formula, which had obstinately withstood its inventor's attempts to explain it away, opened a new door into the atom. First it had led Einstein to an understanding of the true nature of light, then Bohr used it as the foundation of his atomic model, and now de Broglie pushed its interpretation even farther than Einstein had dared. If, as Einstein supposed, a photon is both a wave and a particle, then, de Broglie wondered, is it possible that all material objects share this dual nature as well?

In particular de Broglie suggested that the electron is both particulate and undulatory, an idea that was considered pure fantasy at the time, like the griffin that is both eagle and lion. But de Broglie had no illusions about his proposal. In his thesis he wrote, "It is worth what every hypothesis is, that is, as much as the consequences which can be deduced from it." And in its support he produced an explanation and a prediction.

He imagined the electron as a wave traveling along a narrow canal. If the canal were curved into a complete circle, a continuous wave-train would normally self-destruct by interference: Its peaks would meet up with the troughs of the previous circuit. Only in a few very privileged cases, when the circumference of the circle just happened to equal an integral number of wavelengths, peak would meet peak and trough trough, so that the wave could continue to flow around undiminished. When de Broglie worked out the conditions under which this would be possible, he found that the special orbits that permitted the propagation of waves were precisely the same that were predicted by Bohr's quantization condition. Thus he was able to explain Bohr's successful model of the hydrogen atom by using different assumptions from Bohr, and, in addition, to predict that electrons, being inherently wavelike in nature, would display interference effects like those of light.

This bold speculation was the first major break with the tradition of Democritus. In the Bohr model, electrons were point particles, objects from our own macroscopic experience transposed into the realm of the very small, like miniature grains of sand. The de Broglie electron, on the other hand, is something altogether different; it is a particle and a wave at the same time, a wave-particle, an object that cannot be imagined, much less pictured. So perhaps the Greek artist who designed the ten-drachma coin chose well after all: Bohr's model of the atom was the last one in history that showed unmistakable traces of the legacy of Democritus. With de Broglie our world picture changed irrevocably.

Albert Einstein, who had invented the original wave-particle, the photon, and was criticized for it, was the first to endorse de Broglie's strange idea. In a letter to his French colleague Paul Langevin, dated December 16, 1924, Einstein wrote about de Broglie: "He has lifted a corner of the great veil." To a physicist, higher praise is inconceivable.

And Einstein, as usual, was right. The speculation of the visionary prince sparked an explosion in physics.

3

Quantum Mechanics: The Language of the Atom

In April 1925, a year after de Broglie submitted his doctoral thesis, a liquid-nitrogen bottle blew up in the laboratory of Clinton Davisson, a research scientist at the Bell Telephone Laboratories in New York City. Neither Davisson nor his assistant, Lester Germer, were hurt, but their apparatus was wrecked. In order to investigate the way electrons bounce off metallic surfaces, they had constructed a glass vacuum vessel about the size of a television picture tube that enclosed a beam of electrons aimed at a piece of highly polished nickel, and a little cup, mounted on the end of a movable lever, for catching the reflected electrons. Like the tube in which J. J. Thomson had discovered the electron, the device was a lineal descendant of the "spark in vacuo" apparatus of Franklin's day. The blast broke the container and exposed the target to air rushing into the breach. This in turn ruined the nickel surface by coating it with moisture and oxygen and whatever else was in the atmosphere of New York at the time.

The only remedy was to rebuild the apparatus and to subject the nickel to prolonged, intense heating in an effort to boil off the offending surface contamination. Although this tedious process entailed an annoying delay, it turned out to be the key to a dramatic discovery.

When Davisson and Germer at last resumed their experiment, they were surprised to find that the accident had changed the pattern of scattered electrons. Instead of fanning out in all directions from the initial point of impact with the nickel target as they had before, the electrons now favored certain directions and avoided others entirely. The spray of electrons reflected off the nickel surface looked as though it came from a garden sprinkler with five separate jets, a phenomenon so puzzling that its explanation would require months of further experimentation and a transatlantic trip on which Davisson learned about the work of Louis de Broglie.

What the experimenters didn't realize at first was that their sample had changed. Under the influence of heat, nickel turns crystalline: its atoms, which are normally jumbled together in a random heap, line up like stacked bricks. If electrons were ordinary particles, such a rearrangement would not affect them very much: they would continue to bounce off the nickel target the way tennis balls bounce off a wall, no matter how its bricks happen to be arranged. But if electrons behave like waves, parallel lines of atoms cause them to reinforce each other to form jets in some directions and to fall out of step and cancel each other out in others. As de Broglie had predicted, electron waves display interference patterns and produce striations like those visible between your knuckles when your fingers are lined up in parallel.

Thus Davisson and Germer stumbled upon the first experimental proof of the wave nature of the electron and incidentally found a way to measure its wavelength (the distance from one crest of the wave to the next), which also turned out to agree with de Broglie's prediction. This achievement lent credence to the strange notion of electrons imagined as waves rather than as point masses and thereby opened a new chapter in the taming of the atom. Although both de Broglie and Davisson believed, back in 1925, that electrons have wavelike properties—one man persuaded by theory, the other by experiment—neither one could so much as guess at the nature of those waves. They were like blind men on a beach who hear, and even count, the regular beat of the breakers as they come crashing in, but who have never seen or felt or tasted water.

Clinton Davisson eventually received a Nobel Prize, together with another physicist, Sir J. G. Thomson, who independently demonstrated the wavelike character of electrons by a similar technique. (The name rings a bell: A quarter of a century earlier, J.G.'s father, Sir J. J. Thomson, had earned the Nobel Prize for showing that electrons are particles by measuring their masses. In the good old days of classical physics, phenomena like light, sound, and electricity were thought to be either particulate or undulatory, and the choice between the two options had led to passionate debates. Now, in the strange new world of the quantum, a father and his son could earn high honors by deciding the issue on both sides of the fence.)

The wave-particle duality of electrons lies at the heart of quantum mechanics. What is an electron? Is it a particle or a wave, or is it both, or neither? The same questions had been asked about the photon twenty years earlier, and no satisfactory answer had been found in the interim. The undulatory electron threw physics out of the frying pan of ignorance into the fire of incomprehensibility.

Common sense insists that particles and waves are exclusive categories, so nothing can logically belong to both. Yet the electron evidently does. Nature is telling us to abandon our categories and to invent new ones, instead of trying to force the facts into inappropriate pigeonholes. In this respect the electron is reminiscent of the duck-billed platypus.

When dried skins of that furry, web-footed beast first reached Europe in 1798, they were compared to the mermaids made by artful Chinese charlatans by attaching a fishtail to the body of a monkey, and dimissed as obvious fakes. Later the French naturalist Étienne Geoffroy "proved beyond all possible doubt" that the platypus was not a mammal, because it laid eggs, and the German anatomist Johann Friedrich Blumenbach gave it the name *Ornithorhynchus paradoxus* because it defied existing classifications. But in 1824 its mammary glands were discovered, and forty years later the laying of two eggs by a captive platypus was observed. Eventually taxonomists had to concede that the platypus was both real and new and created a subclass of mammals that lay eggs. Nature does not feel compelled to make use of our preconceived categories, but chooses her own. The electron is neither a particle nor a wave, but something entirely new.

The explosion in Davisson's laboratory turned out to be only a faint precursor of the detonation a month later that would shatter the conceptual foundations of physics and introduce a new view of the electron. In May 1925 the German physicist Werner Heisenberg, then twenty-three years of age, was recuperating from an attack of hay fever on the island of Helgoland in the North Sea. In the placid surroundings of the resort he completed work on a bold new scheme for reconciling the Newtonian laws of classical mechanics with the incompatible finding that energy is discontinuous, that it comes in discrete amounts, as for example in photons and atomic energy levels. Instead of tinkering with the old theories, he took a radical approach, starting with a basic philosophical principle and painstakingly working his way up from there. The result he arrived at surprised him, but from the beginning his instincts told him that it was right and that it represented the doorway to a new and, as he put it, "strangely beautiful picture of reality."

Heisenberg took as his guiding principle the proposition that a theory should not traffic with unverifiable abstractions. He wanted to deal only with measurable quantities and set out to purge the theory of all other elements, which he regarded as metaphysical impediments. He later told Einstein that "the idea of observable quantities was actually taken from his relativity," which had rejected such concepts as absolute speed for the same reason. (If you imagine a universe empty of all matter save one spaceship, then there is no way to define, let alone measure, that ship's speed. But if you introduce another object into this hypothetical universe, say a star, which can serve as a reference point, then the speed of the spaceship with respect to the star can be both defined and measured.) Absolute speed, which Newton had used in his theory of mechanics, is an unmeasurable fiction, so Einstein replaced it with the concept of relative speed. Heisenberg embarked on a program to redesign atomic theory along similar lines.

He began by recognizing that the orbits of electrons in atoms are not observable, much less measurable, and that any attempt to examine an atom with a beam of light or some other probe would utterly disrupt an electron's path. Thus an electron's trajectory is meaningless, and the only observables are the photons emitted or absorbed as the electron leaps from one energy level to another (leaving aside more violent experiments, such as Rutherford's, in which the atom is mangled by an alpha particle). While the electrons of an atom remain at the various discrete energy levels that Bohr had envisioned, its interior is a sealed book, inaccessible to human observation.

Heisenberg concentrated on the messenger photons, noting that every one that interacts with an atom can be characterized by the two energy levels that are involved in its emission (or absorption): the level of the initial state, before the leap (identified by, say, the letter *n*), and that of the final state in which the electron ends up (called *m*). These two labels are the hooks, so to speak, the *only* legitimate handles with which the elusive electron can be grasped.

By following this line of reasoning Heisenberg arrived at the basic elements of his scheme—not mathematical functions, or numbers, but arrays of numbers, like spreadsheets. In each array the rows were labeled according to the initial atomic states, and the columns by the final states. The interpretation of an individual entry in the array corresponding to the position of an electron is something like this: The number at the juncture of the *n*th row and *m*th column represents a measurement of the position of the electron, insofar as it can be deduced from an observation of a photon emitted when the electron leaps from its *n*th to its *m*th level. The photon reveals only a crude approximation of the electron's whereabouts, so this method of describing the atom is much less precise than a planetary model. If the labels *n* and *m* of the atom's energy levels were replaced by the names of cities, the array that represents an electron's position would resemble the chart of intercity distances on a highway map. Thus, if it were ascertained, somehow, that a car is en route from *N*ashville to

*M*ilwaukee, its location could thereby be fixed only within the enormous range of 475 miles—the entry in the intercity mileage chart for row *N* and column *M*.

Other attributes of the electron, besides position, are represented by other arrays. The electron's speed, for example, corresponds to an array that might resemble the chart of intercity driving times, and the array for energy could be analogous to a chart of gasoline consumption between cities. For every one of the electron's features there is such a list of numbers. Each list contains all the information that one can possibly gather about that attribute from direct experimentation.

In order to match the successes of Einstein's photon theory, and Bohr's model of the atom, Heisenberg had to adjust his formalism by fine-tuning it, as it were. His theory automatically led to the *existence* of quantized energy levels—to a staircase rather than a ramp—but without predicting the height of the steps or the *values* of those levels. To complete his theory, he introduced one single new axiom, which is now called Heisenberg's quantum postulate. It contains Planck's constant, which functions like the scale in the corner of a highway map, and thereby allows the theory's qualitative predictions to be made quantitative. The quantum postulate gave his strange new formalism its power and relevance and allowed him to calculate the correct energy levels of the hydrogen atom without invoking a planetary model, or even so much as mentioning electron orbits.

Heisenberg had succeeded in modifying Newtonian mechanics in such a way that quantization and discreteness became part and parcel of the description of the atom from the outset. Bohr had paved the way, but he had had to graft quantization onto Newtonian mechanics as an additional, artificial limitation, whereas Heisenberg's spreadsheets were written in the very language of energy levels. The new formalism offended against one of the most cherished tenets of classical physics, the notion that physical processes are smooth and continuous, that nature, as Newton put it, "does not make jumps." Indeed, Newton had invented calculus as a mathematical technique for expressing the continuity of motion and had called it the method of fluxions, of continuous, flowing changes. Heisenberg, frustrated with the lack of generality of Bohr's patchwork theory, had the youthful audacity to abandon the Newtonian tradition, as well as the artificial restrictions Bohr had imposed on it, and to start afresh.

Since the variables of his theory, such as position, speed, and energy, consisted of entire spreadsheets, rather than individual numbers as in classical mechanics, Heisenberg had to invent new rules for manipulating them. How do you multiply a velocity by a distance, for example, when both are lists of numbers? How do you square such an array or take its square root? Heisenberg had never encountered questions like these in his study of mathematics and had to muddle through on his own. Later, when he returned home to Göttingen, he learned that an array of numbers is called a matrix and that nineteenth-century mathematicians had already worked out the rules for computing with them. But in the isolation of Helgoland, Heisenberg had been forced to reinvent the rudiments of matrix algebra.

The utter novelty of his reasoning, coupled with the unfamiliar mathematical language in which it was couched, prevented its immediate adoption by most theoretical physicists. But there was another, more deeply rooted impediment to its rapid acceptance. The new theory was too austere and forbidding for many people. Heisenberg had made a point of insisting that the structure of the atom is inaccessible to observation and that to describe it in the familiar terms of ordinary sensory experience is illegitimate. "What mental image are we to attach to a matrix?" we ask. "None!" replies Heisenberg's spirit sternly. "Then how can we picture the atom itself?" "Don't try."

Heisenberg did not resolve the enigma of wave-particle duality, the mystery of the platypus. He simply refused to confront it, and ruled out any discussion of words like *particle* and *wave.* Matrices were his ultimate reality. If they did not seem relevant to the ideas we are accustomed to, that was too bad; it was simply the way that nature was arranged.

Heisenberg's outlook on nature was shaped to a large extent by his preoccupation with classical philosophy. In high school, where he had learned Latin and Greek, he had first come across atomism in Plato's *Timaeus,* and throughout his life Plato's idealism, in its geometrical abstractness and pristine purity, served as his guide. In 1975, a year before he died, he remarked in an address to the German Physical Society: "If we wish to compare the results of present-day particle physics with any of the old philosophies, the philosophy of Plato appears to be the most adequate: The particles of modern physics . . . resemble the symmetrical bodies of Plato's philosophy."

His method was to search for the essence behind mere appearance and sensory experience. Matrices, mathematical abstractions like Plato's ideal solids, were more real to him than material bodies. Although his theory seemed to differ sharply from the concrete planetary model it replaced, Heisenberg's purpose had much in common with Bohr's. He, too, was searching for eternal truths and never regarded his own early contribution as more than provisional. Not surprisingly Bohr was one of the first to appreciate the significance of the quantum revolution: In August 1925, just a few weeks after the appearance of Heisenberg's paper, Bohr tried to explain it to the Scandinavian Mathematical Congress, meeting in Copenhagen, and hailed it as an outstanding achievement. It is likely that few in the audience understood what he was talking about.

Not by accident, an older physicist, with less of the brashness of youth and closer ties to everyday reality, soon provided relief from Heisenberg's intellectual aloofness. In early 1926 the Austrian physicist Erwin Schrödinger, who was thirty-eight years old at the time, independently proposed another formulation of the same theory. Although his concepts and language bear no resemblance to those of Heisenberg, he was able to show that the two versions were logically equivalent. Since Schrödinger's reasoning is much more accessible, it has largely replaced the earlier scheme for routine calculations. But the disquieting fact remains that Schrödinger's picture of the atom can be recast into the matrix mechanics of Heisenberg, which in turn is grounded in the principle that an accurate mental image of an atom cannot be drawn.

If Heisenberg's description of an electron within an atom resembles intercity travel charts, Schrödinger's is like the map itself. It conveys a picture, not just a set of numbers. It is analog, not digital. The metaphor can be pushed a little farther: Think of a highway map with a great number of cities and towns but only straight sections of road between them and no other details. Such a map is logically and mathematically equivalent to a complete matrix of distances. One can easily determine distances from the map, and conversely, armed with a little geometry and a lot of patience, one could reconstruct the entire map from a complete chart of the distances. In this light it is no wonder that most physicists prefer Schrödinger's quantum theory to Heisenberg's. Who wants to navigate across unfamiliar territory with a mileage chart instead of a map?

Schrödinger came to quantum theory by way of de Broglie's hypothesis, though initially as a skeptic. His first reaction to the suggestion that an electron is a wave was ridicule—"rubbish" he is said to have called it—but others prevailed upon him to take it seriously. So he set out to answer a simple question that de Broglie's hypothesis had raised: If the electron is a wave, what is its wave equation? In classical physics a wave equation is a mathematical relationship that describes all waves, be they ocean waves, sound waves, light waves, or the ripples on a flag. It always has the same general form, but differs in detail from case to case. In the case of the electron the wave equation was unknown, but Schrödinger, proceeding by guesswork and a lot of trial and error, succeeded in finding it. Today this formula bears his name and constitutes the principal mathematical tool for describing the atom. It is only a mild exaggeration to say that theoretical atomic physics is the study of solutions to the Schrödinger equation.

But where does the discreteness of energy levels come from? Why does light come in bundles rather than a steady stream? Where is quantization in this picture? Ocean waves breaking on the beach are perfect examples of continuous flux and reflux, without a hint of discreteness. There are rollers with low frequencies and ripples with high frequencies and all possible frequencies in between. How can a wave equation account for the quantization of the energy of electrons?

The answer is that there is a subtle but profound difference between an ocean wave and an electron in an atom: The former is free to spread out across the entire vast ocean, while the latter is confined by electrical attraction to the vicinity of the nucleus. Waves that are confined to a limited space are quantized, even in the ordinary macroscopic world. This fact is the secret of how music is made. When a string is made immobile at both ends, as it is in a piano, it can vibrate with only one specific frequency (and its harmonics) corresponding to one specific pitch (and its overtones). The same is true for the column of air in a flute and the membrane of a drum. Music is made possible by the quantization of confined waves, and so are atoms. The Schrödinger wave equation at last furnished a natural explanation for the discreteness of atomic energy levels first proposed by Bohr. The metaphor for the atom had shifted, from the solar system to the piano.

Thus half the battle was won: The electron's wave nature had been captured mathematically, while the other half, its particulateness, remained a mystery. The connection between the wave equation and the electron's particle properties, when it was made, turned out to be related to the interpretation of the theory. Schrödinger's wave equation has a solution, a mathematical expression that is usually denoted by the Greek letter psi, and is called the wave function. In classical physics the solution of a wave equation has a perfectly well-defined meaning that depends on the case under discussion. It may be the height of an ocean wave, the pressure of air in a sound wave, the strength of the electric field in a light wave, or the displacement of the ground in a seismic wave. But what is the physical significance of the quantum mechanical wave function? What does psi correspond to in an actual atom? Credit for answering this crucial question, and thereby completing the description of the electron, came to its discoverer late in life. It was Max Born, Heisenberg's professor in Göttingen, who gave meaning to the wave function soon after it was introduced by Schrödinger. But while Heisenberg and Schrödinger were rewarded with Nobel Prizes in the early 1930s, Born was not—a fact that by his own admission hurt him very much at the time. When the slight was finally righted in 1954, twenty-eight years late, Born speculated that the delay might have been caused by the opposition to his idea by many of the architects of the quantum theory. Planck, Einstein, de Broglie, and even Schrödinger himself expressed serious philosophical objections to Born's novel interpretation of psi. But Born's view has prevailed and is commonly accepted today.

In 1926 he made the startling suggestion that the quantum mechanical wave function must be interpreted differently from all its classical counterparts. Rather than measuring something that *is* there, as the other wave functions do, it refers to something that *might be* there. The wave function determines the probability of finding the electron at a given spot. Since detectors, such as photographic plates, fluorescent screens, and electronic counters, record individual particles and not waves, the interpretation of the wave function as probability provides a link between the two seemingly incompatible facets of the electron's dual nature.

And so the platypus was found. Today the electron's strange attributes have all been described in the mathematical language of theoretical physics. But the translation of that language into ordinary English sounds curiously Aristotelian: The wave function measures potentiality, not actuality. This idea can be illustrated with reference to a planned vacation. Suppose that you live in St. Louis, and that although you are not particular about the direction in which you travel, you don't want to go beyond, say, a thousand miles away. If every spot in the country is assigned a number to represent the probability of becoming your destination, St. Louis will get zero, and so will far-off San Francisco. Around St. Louis the numbers will be arranged in concentric circles, like a rifle target, rising and then dropping off again until they reach zero beyond a radius of a thousand miles. This is a map of your private vacation probability; it depicts your wave function.

Such a map contains a lot of information, but fails in one

crucial respect: It does not show you where you are actually going to spend your vacation. Once you buy your ticket, pack your bag, and set off, the map becomes irrelevant, just as a wave function becomes irrelevant after the electron has actually been detected. But before that decisive intervention occurs, the wave function contains all the information that can possibly be obtained about the electron's future behavior.

The wave function and the result of an experiment usually look very different from each other; the wave function often displays a much higher degree of symmetry than a measurement. For example, your vacation map resembles a series of circles around the city of St. Louis, but when you take your actual vacation in a specific place, say Chicago or New Orleans, the map that records your location bears no trace of that symmetry. Analogously the wave function of a hydrogen atom in its lowest energy level is spherically symmetric—a ball of probability around the nucleus. But when the electron in a hydrogen atom is somehow captured, its location is a point and reveals no symmetry at all.

Although the vacation analogy is helpful, it misses the true mystery of quantum mechanics. Compared to the map of vacation probabilities, which has a simple but abstract meaning, the wave function has much more physical significance. In fact the electron is in a sense spread out, as though it were a real wave whose magnitude is described by the psi function. And yet Born's interpretation of Schrödinger's wave function assumes that the electron is a point particle, not a wave. The seeming contradiction between the description of the electron as a minute particle and as a spread-out wave is the modern version of the old wave-particle duality, and remains enigmatic. Richard Feynman called it "the *only* mystery," and added "we cannot make the mystery go away by 'explaining' how it works." But we can describe it in concrete terms.

To see that an electron really is spread out, that it can in some sense occupy several places at once, we return, at Feynman's suggestion, to an interference experiment, albeit an idealized one that is conceptually much simpler than Davisson and

Germer's investigation of reflection from nickel. The experiment happens to be the one used by the English physician, physicist, physiologist, and linguist Thomas Young in 1803 to demonstrate the wave nature of light. Whenever the subject of waves comes up, whether it is with reference to light, radio signals, sound, or electrons, Young's experiment is the first one physicists turn to. It has been repeated in countless forms, from high school demonstrations with water waves to the most sophisticated investigations at billion-dollar particle-accelerator laboratories. Its simplicity and cogency make it one of the classic experiments in physics, an intellectual monument of Western culture as lasting and inspiring in its sphere as a sculpture by Michelangelo and a sonnet by Shakespeare are in theirs. Neither time nor familiarity impair the freshness and fascination of Young's experiment, any more than they erode the grandeur of the greatest works of art and literature. And yet its mechanism is basic and unpretentious.

Young pricked two pinholes, side by side and a fraction of an inch apart, in his window curtain and examined the sunlight that streamed through them onto the opposite wall of his room. What he saw there was not a pair of spots, but something much more intriguing. He found an oval patch of light that was interrupted by a series of equally spaced vertical dark stripes. Young's interpretation was that waves coming through the two pinholes were alternately in step and out of step with each other, proving categorically that light is wavy. More important, no particle model of light could explain the phenomenon: Two simultaneous bursts of lead pellets from a double-barreled shotgun don't cancel each other out.

When the same experiment is performed with electrons instead of light, the technical details necessarily differ, but the conclusion is the same. The result is similar to the observation by Davisson and Germer that electrons display dark interference stripes like light. The electronic version of Young's experiment thus seems to achieve nothing more than a corroboration of the wave nature of electrons. But very careful thought leads to a disturbing conclusion.

Imagine that the initial electron beam is pathetically feeble.

For precision assume that only one single electron arrives at the pinholes every minute, passes through, and makes a dot on a photographic plate on the wall. The dots produced by the first ten electrons produce a fairly nondescript, random-looking pattern. But after many hours, and hundreds of electrons, an image emerges clear and sharp: an elongated patch crossed by a series of regularly spaced dark lines. And therein is hidden the mystery of quantum theory in its quintessential form.

Consider a single electron that passes through one of the pinholes and lands on the wall to make a dot. Somehow this particular electron is prevented from hitting those places that will become dark bands many hours later. Something—an unknown force, an unseen influence, some hidden predisposition—prevents the electron from arriving at those special, well-defined forbidden locations.

Such a force cannot be related to other electrons, because those are far away in both space and time; the next one is not due to arrive for another minute, which might as well be a year. Other electrons are simply irrelevant to the whole discussion. Nor could the curtain itself be capable of steering the electron in the appropriate direction, because covering up either one of the two holes causes the dark stripes to disappear. In other words an electron passing through a single hole is perfectly capable of striking the places that are forbidden when the two holes are open.

The mystery vanishes instantly when the initial beam consists of waves. The light waves of Young's original experiment, for example, produce the black stripes by passing through both holes simultaneously. If we adopt a similar explanation for electrons, we must conclude that the electron passed through both holes at once, even though every time a single electron is detected by a photographic emulsion, it is a pointlike particle, much smaller than the distance between the holes, or even the diameter of a single hole. And yet this little speck manages to pass through two holes that are separated by what to the electron is an enormous distance. The electron somehow manages to do what only spirits are supposed to be able to accomplish: It occupies two places at once.

Quantum mechanics relieves this perplexing conclusion a little bit. Born's interpretation of the Schrödinger wave function states that only the electron's probability of being found, not its material self, behaves like a wave. But this clarification doesn't entirely resolve the paradox. If the wave function is sufficiently powerful to direct the electron's path away from the forbidden zones, it must have actual, physical content. Considering the analogy of potential vacation plans, the map only *seems* to direct you away from St. Louis and from the distant coasts, but it does not prevent you from choosing one of those less likely places. The real mechanism in control of the vacation is simply your volition—the map merely reflects your wishes and has no power of its own. In the case of the electron, on the other hand, volition is absent. The electron is controlled by real physical influences, and the wave function describes their effects. The wave function is the probability map of the electron's potential positions, and it behaves as if it were a real fluid.

That, in a nutshell, is the mystery of the quantum: When an electron is observed, it is a particle, but between observations its map of potentiality spreads out like a wave. Compared to the electron, even a platypus is banal.

Young's experiment only illustrates the quantum mechanical character of a single electron—in particular its wave-particle duality—but there is another property of electrons that refers to interactions rather than to individuals. When two electrons meet, they repeal each other electrically, and each one has all the peculiarities of a wave-particle, but they also exhibit an additional quantum effect that has no counterpart in the ordinary, macroscopic world. Quantum theory teaches that at the fundamental level of matter, just as in the human realm, it is not enough to describe individuals—the relationships between individuals must also be considered.

The roots of these quantum mechanical relationships reach all the way back to Melissus, the pre-Socratic philosopher who had inspired the atomic hypothesis of Leucippus and Democritus. His maxim, But if there are Many, they must each have the character of the One, could be considered to be a forerunner of the modern principle of indistinguishability, according to which all electrons (as well as all other particles of the same species) are indistinguishable from each other. For a fundamental principle of theoretical physics, that sounds pretty innocuous. Aren't all ten-drachma pieces the same, too, and all pellets of lead shot, and all raindrops? Isn't indistinguishability a commonplace concept? In fact it is not. In the world of everyday experience no two objects are truly identical. Examined with sufficient magnification, all coins and all pellets, like all grains of sand and all snowflakes, differ slightly in detail. The concept of sameness, once it is taken beyond the superficial level, is strictly an atomic phenomenon, and quantum mechanics derives unexpectedly powerful consequences from it.

The principle of indistinguishability imposes a discipline on systems of many particles that is comparable to two extreme forms of social contract. Certain kinds of particles, including photons and alpha particles, are required to behave in perfect harmony, each going through the same motions as all its peers—a kind of communism pushed to its limits. Other particles, including electrons and hydrogen nuclei, called protons, behave in the opposite way: No two are ever allowed to be in the same state of motion—a type of capitalist individualism run amuck. If everyone is the same, or if everyone is different, it is impossible to single out anyone from a crowd, and that is the essence of indistinguishability. (The explanation of why photons happen to fall into one camp and electrons into the other did not emerge until the 1940s, when quantum theory, special relativity, and electromagnetic theory were fused into the modern framework of quantum electrodynamics.)

The rule that electrons must all behave differently has a profound consequence for the structure of atoms. Called the exclusion principle, it states that no two electrons can have the same energy. The exclusion principle resolves a fundamental difficulty related to the architecture of matter. When the atom was dissected into a positive nucleus and a cloud of negative electrons, the question arose, Why doesn't it collapse as its electrons are pulled into the nucleus? The Schrödinger equation furnished part of the answer because it did not have a solution corresponding to an electron being found to occupy the center of the atom, any more than a drumhead is capable of vibrating only dead center and nowhere else, but that wasn't the end of the story.

In an atom like helium, with its two positive charges in the center, why don't both electrons simply jump down into the smallest possible orbit, which corresponds to the energy level of hydrogen? If they did, the structures of helium and hydrogen would be almost identical, their enormous physical difference—hydrogen is explosive while helium is inert—would be erased, and the universe would be a very different place. The reason electrons stay on separate energy levels of the atom is very simple: They have to. The exclusion principle constrains each additional electron in the outer shell of an atom to seek out a higher energy level, and the wave function locates it farther from the nucleus than the one before it. Going beyond hydrogen and helium, the exclusion principle accounts for the diversity of the chemical elements.

The periodic table of elements is ordered according to the number of positive charges in the nucleus. If one imagines beginning with hydrogen and adding positive charges to the nucleus one by one, and electrons to the outer shell of the atom one by one to ensure electrical neutrality, then each electron has to be in a larger orbit than its predecessor and carry more energy. (In real atoms this simple recipe is somewhat more complicated, chiefly on account of the magnetic properties of electrons, but the essential idea is the same.) In this way the entire periodic table of elements is built up. If electrons and nuclei are the bricks of the material world, wave-particle duality applies to individual bricks, and the exclusion principle regulates how they are stacked up.

It all sounds so neat and cogent. But looking back at the explanation, some questions remain. How can an electron be in two places at once? How can one electron know what the other one is doing, even at a great distance, so that it can obey the exclusion principle and avoid doing the same thing? Have physicists given up the goal of understanding the world as it really is and reverted to attributing unexplainable phenomena, like the exclusion principle, to ghosts and spirits? Or have we become content to describe atoms by psi functions that obey arbitrary rules without worrying about what they mean?

It bothered Einstein very much, this psi function. Although he had invented the first wave-particle, the photon, and recognized the significance of de Broglie's hypothesis earlier than most of his colleagues, although he had helped Schrödinger work out the logical development of his equation and contributed in innumerable ways to the quantum-theoretical description of atomic physics, he could not bring himself to remain satisfied with Born's interpretation of psi as a measure of probability. Somehow, he felt, God must know for sure where the electron is and where it is going. And if God knows, we should be able to figure it out too. "God does not play dice," Einstein exclaimed, and strived to reintroduce the word *reality,* which Heisenberg had banished. Quantum mechanics, he said, is an "incomplete description of physical reality."

Prince Louis de Broglie went farther. Impressed by the successes of quantum mechanics, he adopted the prevailing interpretation of the wave function for several years, but then his philosophical scruples reasserted themselves. He wrote that psi is "a simple representation of probabilities that leads to a great number of exact predictions but does not give any understandable representation of the coexistence between waves and particles. . . . I have been convinced again that we have to come back to the idea that the particle is a very small object which moves along a trajectory." When he died in 1987, two generations after he had set off a revolution that irreversibly changed the scientific world, his intuition had carried him back to the physics of the seventeenth century.

Erwin Schrödinger tended in the opposite direction. The explanatory power of his own equation persuaded him that ultimately the world really is undulatory, that at the most fundamental level it consists of waves, not particles. To Schrödinger, psi was a description of the actual electron, a map, so to speak, of its whereabouts, just as a weather map is a representation of the actual location of clouds. Indeed, many physics textbooks include pictures of the hydrogen atom made by assigning different depths of shading to different values of psi (or its square): in those places where the calculated value of psi, which is to say the probability of finding an electron, is largest, the image is darkest, and where psi has a low value, the image is faint. The picture of the hydrogen atom's electron in its lowest possible energy state, for example, looks like a cloudy crystal ball centered on the nucleus, opaque in the center and growing increasingly translucent toward the edge. Such images are sometimes called charge clouds, because they graphically represent the distribution of the electron's electrical charge in space. Where they are most dense, the electron with its negative charge is most likely to be found. It is tempting to assume that the picture of a charge cloud is a realistic depiction of the atom and should replace Bohr's planetary model as icon. But this assumption is wrong.

The first problem, as Max Born never tired of telling Schrödinger, is that such realistic interpretations of the wave function are not tenable for atoms with more than one electron. The difficulty is not one of detail or mathematical complication; it is fundamental. In the Schrödinger equation each electron occupies its own three-dimensional space. The lithium atom with its three electrons, for example, is described by a psi function in nine-dimensional space, which is perfectly suitable for making predictions about what happens in a laboratory experiment, but it is obviously not the space we live in. And if lithium isn't bad enough, who could comprehend the eighty-four-dimensional space of table salt?

The second problem with charge clouds is that they represent, in the end, potentiality, not objective reality. A map of probabilities cannot possibly represent a real object. For example, in Young's experiment performed with a single electron, the wave function, not the electron, squeezes through two pinholes and then fans out in ripples and striations to the distant screen. If one were to adopt that as a real portrait of an electron, all connection with common sense would be severed.

The atom is difficult to imagine. The vocabulary of quantum mechanics, which describes electrons with great precision, is not easily translated into the words and images of everyday life. Our lack of understanding is not on the technical level. On the contrary, the success of quantum mechanics, measured by the accuracy of the quantitative predictions it makes, is unprecedented in physical science. The fault, rather, is in the interpretation of the theoretical concepts; we understand the substance of quantum mechanics, but not its meaning.

This is the dilemma we have come to, step by step, since Democritus first introduced invisible philosophical atoms to explain the real, visible world. He taught us to understand the properties of matter in terms of those minute kernels, and his successors through the centuries—Lucretius, Newton, Bernoulli, Dalton, and all the others—perfected the image of the atom. Thus we learned to imagine the atom on the basis of theory and speculation far removed from direct experience. Today the gulf between what we know from theory and what we see with our eyes has been pushed downward into the interior of the atom, but it has not been bridged. On the contrary, if anything, it has widened. Tom Stoppard's quip about quantum mechanics below, classical physics above, and metaphysics in between, describes the problem perfectly.

From below, quantum mechanics reveals the structure of the atom with increasing precision, but fails to furnish a satisfactory mental image of its interior. From above, new and increasingly powerful instruments reveal the exterior, and thus elevate it to the sphere of classical physics. Instead of contracting ourselves to atomic dimensions, as Thomas Harriot had suggested, we have succeeded in magnifying atoms to our size so that we can see and manipulate them. And what we see is not so different from what we had imagined all along.

The Present

4

Images of Atoms

In March 1850, Friedrich August Kekulé, a young student of architecture, appeared before a grand jury in Darmstadt, Germany, to help clear up the circumstances of the death of his neighbor, Countess Görlitz. Her charred body had been found in an otherwise undamaged room a few weeks earlier, and the cause of death was assumed to be spontaneous combustion brought on by excessive drinking.

Among the other witnesses called was the organic chemist Justus von Liebig, who testified that spontaneous combustion of human tissue is physically impossible; alcohol would poison a person long before it could raise the flammability of the body by an appreciable amount. Kekulé was asked about one of Görlitz's servants, a man called Stauff, who had been caught a few days before the trial selling stolen goods, including a gold ring in the shape of two intertwined snakes biting their own tails, the alchemical symbol for the unity and variability of matter. The ring, Kekulé told the court, had been the countess's talisman. This, in conjunction with other evidence, convinced the jury that Stauff was guilty, and he was convicted of murder.

The trial changed Kekulé's life as well. Professor von Liebig's testimony made such a deep impression upon him that he switched his studies from architecture to chemistry and chose von Liebig as his mentor. Even the incriminating ring left a lasting impression, for it was the serpent symbol that bubbled up from Kekulé's subconscious fifteen years later and led him to his famous solution of the problem of the structure of the benzene molecule.

Given that Kekulé's first interest had been architecture, which deals with the arrangement of matter in space, it is not surprising that his greatest contribution to science was in steering chemistry away from the mere identification of the atoms that form molecules, and toward the investigation of how these atoms are arranged. Methane, or common marsh gas, was the first molecule he tackled in this new way—a simple task compared with the more complex benzene, which came much later. Judging from the weight and chemical properties of methane, Kekulé knew that its molecule contained one carbon and four hydrogen atoms. But there are myriad ways in which the five atoms might combine, ranging from a ball-and-chain structure to a compact clump. Each arrangement would have different consequences for the ease or difficulty with which methane could combine with other atoms to build more complicated molecules. Kekulé was thoroughly familiar with the chemical properties of marsh gas, and reasoning on the basis of those, he tried different configurations of atoms, but to no avail. The solution, he claimed, finally came to him in a dream. He told my great-grandfather, Adolf von Baeyer, who was his first research assistant, how it happened.

It was during the early 1850s, when Kekulé was living in London. Late one night he was returning by omnibus from a visit to a friend across town. As usual he climbed up the narrow staircase in the rear of the horse-drawn carriage and sat on one of the benches on the roof. The warm evening air and the empty streets induced a sort of reverie, and he had a vision. Atoms sprouted arms, whirled about, and seemed to reach out to each other, and by the time his waking dream had come to an end, Kekulé had deciphered the structure of methane: a carbon atom, from which four arms grow in the form of a cross, with a hydrogen atom at the end of each. Chemists later refined this picture into a three-dimensional model in which the carbon atom is surrounded by equidistant hydrogen atoms at the corners of a tetrahedron, or triangular pyramid, but Kekulé's revelation was essentially correct.

In the hands of Kekulé and his colleagues throughout the world, structural chemistry, which relates the chemical properties of molecules to their spatial structures, proceeded apace. Once they had developed a model of the basic ingredient methane, they explored the ways in which other atoms could be attached to it. They spliced, merged, and recombined, eventually synthesizing dozens of compounds, some already known, others never observed before. From today's perspective the most astonishing aspect of this research is that it was carried out with test tubes and retorts containing liquids, vapors, and salts, that its chief diagnostic tools besides the scale and the thermometer were the eye and the tongue, and that it was conducted long before anyone had seen an actual molecule—before many scientists had even accepted the concept of atoms as physically real objects. It is no wonder, then, that structural chemistry did not meet with universal acclaim. "Not far removed from belief in witchcraft and spirit rapping" is how the German chemist Hermann Kolbe characterized such ideas.

But Kekulé was not dissuaded; in fact he encountered no significant obstacles until he attempted to decipher the structure of benzene. If the benzene molecule is thought of as an open chain of six carbon atoms, each with one hydrogen atom attached to it, then it should not be too different from similar chains having seven carbons, or five. But benzene, as a chemical, seemed more stable and self-contained; it did not share the propensity of carbon chains to capture additional atoms at their ends, and somehow, Kekulé felt, this difference should be reflected in its molecular structure.

In the midst of his puzzlement Kekulé had another dream. "One night," he later wrote, "I turned my chair to the fire and sank into a doze." There appeared before his eyes a vision of atoms that danced and gamboled as before, occasionally joining together and turning and twisting in snakelike motion. Then, suddenly, "one of the serpents caught its own tail and the ring thus formed whirled exasperatingly before my eyes. I awoke as by lightning, and spent the rest of the night working out the logical consequences of the hypothesis." Kekulé proposed that the benzene molecule is a hexagonal ring of carbon atoms from which six hydrogen atoms dangle like charms from a bracelet.

The ring theory overcame the difficulty of the open-chain models and afforded to benzene a special status. Although it took some time for Kekulé's hypothesis to gain acceptance, it eventually launched an entire branch of structural chemistry—the study of ring-shaped compounds. For more than a hundred years Kekulé's model has been considered an established fact, which is remarkable, given that until very recently no one had so much as glimpsed an actual benzene molecule.

Kekulé, like Thomas Harriot in the seventeenth century and Ernest Rutherford at the beginning of the twentieth, had the ability to see atoms with his inner eye, but the rest of the world got its first view of the benzene molecule in 1988. That year the June-sixth issue of the journal *Physical Review Letters* carried a black-and-white picture of what looked like a tray of lumpy doughnuts, with tiny dark smudges where the holes should have been. In fact it wasn't a picture of doughnuts at all but an image of benzene molecules attached to the surface of a strip of rhodium metal; each pierced blob represented a single, ring-shaped molecule of the organic chemical. The diameter of the molecule was reported to be approximately a billionth of a meter, so that a million of them, lined up in a row, would stretch across a period on this page. The image was produced with a new device, the scanning tunneling microscope, or STM, by scientists at IBM's Almaden Research Center, in San Jose, California.

Inasmuch as chemists have known about the benzene ring for generations, the IBM microscopic image, or micrograph, was an anticlimax. Even the article that accompanied the picture glossed over the topic of the molecule's structure, emphasizing instead the promise the technique holds for the study of chemical processes on surfaces, such as corrosion and adhesion. If anything, by confirming Kekulé's model, the image inadvertently demonstrated the power of the human mind, unaided by modern machinery, to synthesize vast amounts of complex chemical information and to render the result in visual terms.

From another point of view, however, the micrograph raised an interesting question. Contrary to appearances, it was not an actual photograph but a computer reconstruction (like a CAT scan of the brain) based on measurements of the electric current that flowed through the tip of a needle as it passed across the molecule's surface. A hidden chain of readings, calculations, and interpretations stood between the sample and the final image. Since Kekulé's model, too, depended on the theoretical interpretation of diverse chemical observations, we are led to wonder about the relationship between the two ways of seeing: To what extent does the imagination furnish an accurate picture of the world? And conversely, how much of the micrograph was, in fact, imagination?

Surprisingly, the making of an STM image is easier to understand than the delicate chemical process that leads to an ordinary photograph. For the light waves used in ordinary microscopes the STM substitutes electron waves, and the specific phenomenon it exploits is known in optics as frustrated total internal reflection. Internal reflection is responsible for the appearance of a rainbow when sunlight enters a raindrop in a cloud, traverses the drop, bounces off the far surface, and finally leaves the drop in the opposite direction. If a light beam happens to strike an internal surface of a drop of water or a piece of glass at a very shallow grazing angle, the internal reflection becomes total, and no light can escape from the interior. Total internal reflection underlies telephone transmission by optical fibers: Since the signal travels along the axis of the fiber, it always encounters the outer walls at shallow angles and therefore rattles back and forth in its narrow channel, like a bowling ball in the gutter, never escaping even after thousands of miles of travel.

There is one way to overcome total internal reflection and to coax light out of its confining glass, even if it arrives at the internal surface at a glancing angle. Imagine that a second piece of glass is brought very close to the first one, almost touching the point where the beam is reflected. Now some of the light will hop across the gap and proceed, as if nothing had happened, into the second piece of glass. Photons, when they arrive at an interior surface, extend their electromagnetic fields—the ethereal stuff of light—like miniature antennae out into the void beyond the boundary, and, encountering no solid substance, turn back into the glass. But if the fields encounter a bit of glass out there, on the other side of a narrow, empty chasm, they generate new photons in the second piece of glass and thus re-create the beam. The width of the gap cannot be much greater than the wavelength of the light, which corresponds to a fraction of the thickness of a fine hair, but the effect is real and easily demonstrated experimentally. As the gap widens, the effectiveness of the "antennae" diminishes rapidly, and the amount of light that escapes the first piece of glass decreases correspondingly.

All of these facts were known to the German physicist Gerd Binnig when he was putting the finishing touches on his Ph.D. dissertation at the Wolfgang Goethe University in Frankfurt in the fall of 1978. He also knew that electrons, having wavelike characteristics, can cross narrow gaps just like light. For example, electrons move around freely inside metallic conductors, such as copper wires, and when they encounter a surface, they are internally reflected back into the metal. (That's what prevents electricity from leaking out of household wires.) But if a second conductor is brought sufficiently close to the first one, the electron can jump across the intervening gap. Because such a gap normally acts as an impenetrable barrier, the word *tunneling* is used to describe the process, even though the barrier through which electrons tunnel consists not of a material obstacle but of its exact opposite, a vacuum. The phenomenon of tunneling is similar for photons and electrons, except in scale; where the gap in the case of light must be narrower than a human hair, for electricity it has to be a thousand times smaller. The width of the gap that allows electrons to tunnel through it is measured in atomic dimensions.

One day Binnig happened to be talking to the Swiss physicist Heinrich Rohrer from the IBM Zurich Laboratory about new ways of exploring the surfaces of materials, especially materials of interest to the manufacturers of computer chips. Binnig suggested that the tunneling of electrons through a vacuum might prove to be useful for surface studies, because minute electrical currents can be measured accurately and because the effect is sensitive to microscopic dimensions. If a negatively charged needlelike probe were brought near a metal surface, electrons could tunnel out of it and thereby reveal the distance between probe and surface. Rohrer liked the idea and invited Binnig to join him in Switzerland to begin the difficult task of translating this hunch into a new instrument.

The collaboration between Rohrer and Binnig, fourteen years his junior, worked out splendidly. They shared a fundamental interest in that both had worked throughout their careers on the problem of superconductivity, and the scanning tunneling microscope was inspired by their desire to understand that puzzling phenomenon. In appearance the two men represent the new breed of European scientist, light-years removed from the stuffy professors of the era of J. J. Thomson and Max Planck. Their casual, open-necked shirts and loose trousers, Rohrer's trim white beard and wire glasses, and Binnig's handsome features combine to make them look like a father-and-son team on an American TV program. Actually they are brilliant, hard-driving scientists with a remarkable, tireless appetite for work.

The chief obstacle they faced in their investigation was the size of the probe. A needle that maps details of a surface must be smaller than those details, for the same reason that the grooves of a phonograph record can't be felt with a finger. In 1978 the sharpest tips that could be manufactured were hundreds of atomic diameters thick, so the new technique seemed incapable of registering atoms. Then, on January 5, 1979, Binnig figured out a way to overcome this limitation.

His laboratory-notebook entry for that day contains a simple drawing that represents a close-up of the tip of a dull, stubby needle touching a flat metallic surface: a semicircle touching a straight line. (Geometrically this configuration is called an osculation, or kiss.) On either side of the point of contact the distance between the surface and the tip of the needle increases in wedge-shaped fashion. According to the cryptic notes to the side of the drawing, Binnig suddenly realized that tunneling diminishes so rapidly with distance that the current between the needle and the sample, which is of course strongest at the point of contact, drops off sharply away from that point. The electrons' wave functions, acting as "antennae," just don't reach out very far. In fact, as the gap widens by a single atomic diameter, the tunneling current drops by an astonishing factor of a thousand. This means in turn that only a tiny patch immediately surrounding the kissing point, not the entire tip, conducts current. (The conclusion remains valid even when the tip is pulled back a little so that it doesn't actually touch.) Electrons will not leave the stubby needle except through that little patch at its extreme tip. In effect the needle is sharper than it looks—about forty times sharper, according to Binnig's estimate.

Below the drawing an exultant fat arrow points to the remark "Microscope for metal and semiconductor surfaces." Binnig had found the key to the invention that would lead to the 1986 Nobel Prize for him and Heinrich Rohrer a scant seven years later.

The device they developed is commercially available today in sleek models no bigger than a teacup. In the beginning it was much larger and clumsier, mainly because of the elaborate scaffolds of springs and shock absorbers that were designed to prevent stray vibrations, caused by seismic shocks as well as passing footsteps and nearby trains, from overwhelming the signal. When you are exploring terrain at the atomic scale, unwanted motions of even the minutest dimensions must be avoided. These technical difficulties were quickly overcome, and the size and complexity of the apparatus diminished, just as the size of calculators plummeted from desk-size to postcard-size in the span of three decades. The heart of the scanning tunneling microscope, however, has remained virtually unchanged since its invention.

The needle, or stylus, which is made of tungsten, comes to as fine a point as is technically feasible. In fact it turns out that the tip usually consists of a protruding imperfection that comprises only a few atoms, and is therefore capable of even finer imaging than Binnig had originally dared to hope for. Guiding the needle across the surface under study is a delicate task because the distances involved are too small for direct manual control, even with the best optical microscope. In some models this dilemma is overcome by mounting the stylus on three toothpick legs made of quartz crystal, which have the intriguing property of changing length under the influence of an externally applied voltage. Thus by judicious adjustment of the voltage on each leg the needle can be nudged in any direction.

In operation the needle's tip is brought close to the metallic surface, and an electrical potential of a fraction of a volt is applied across the gap to stimulate the tunneling of electrons. The needle is then systematically directed across the sample. When the probe encounters a bump on the surface, the width of the gap diminishes, with the result that the tunneling current rises. The probe is then automatically pulled back, away from the sample, until the current drops to its former value, and the motion of the needle is recorded by a computer. Conversely when the current falls on account of a dip, the probe is automatically eased in closer to the surface. In this way the entire surface is explored, with the tunneling current, and hence the distance between tip and sample kept at a constant value.

The coordinates of the needle, including its regular to-and-fro sweep across the sample as well as its vertical variations, are recorded. When all the data are in, the computer reconstructs a contour map of the surface, in the way that a landscape might be mapped out by an airplane flying at a constant height above the ground, climbing and dipping as the territory undulates beneath it. Finally the completed contour map is displayed on a monitor. That is how benzene molecules stuck to rhodium metal became visible for the first time in 1988.

Actually images of molecules have been available since the second decade of the century, when X-rays began to be applied to the exploration of the structure of matter, but they are pictures only in a very special sense. X-rays bounce off atoms the way sunlight bounces off raindrops, and what is recorded in an X-ray image resembles a rainbow. Just as the theory of the rainbow can be used to reveal, indirectly, the shapes and sizes of individual drops, the theoretical analysis of X-ray pictures gives information about the arrangement of atoms in molecules. But a rainbow is not a picture of a raindrop, nor is an X-ray picture an image of a molecule. The value of the scanning tunneling microscope is that, for the first time, it reveals the architecture of individual molecules one at a time.

The images produced by the STM are still not pictures in the conventional sense, because they depend on electrical currents, not light, and therefore represent a new way of seeing. The computer records the strength of the tunneling current, which is determined by two influences, neither of which is associated with vision. The first is the proximity of the tip of the probe to the surface under investigation: The narrower the gap, the stronger the current. This relationship is the crux of the operation of the STM, for it alone reveals the atomic structure of the sample. But the second effect, caused by the sample's electrical charge, cannot be disentangled from the first. When the probe is negative and the surface positive, electrons ooze from the needle down into the sample. (When the polarity is reversed, electrons can travel in the opposite direction just as well.) Suppose, now, that lying on the surface of the sample, just under the negative tip, there is a large atom. The atom as a whole is neutral, but its charges are separated, with the nucleus charged positively and the outer shell negatively. The charge on the surface of the atom repels the tunneling electrons and decreases their flow across the gap, so that the STM registers readings identical to those caused by an indentation in the surface, which also inhibits the tunneling current. On the other hand, an excess positive charge on the surface under investigation enhances the current and is automatically recorded in the computer as a bulge. In this way electrical and structural features are entwined.

Sometimes this problem leads to confusion between dips and bulges. In 1987 a team of physicists at the IBM Thomas J. Watson Research Center in Yorktown Heights, New York, decided to explore the oxidation, or rusting, of the surface of gallium arsenide, a compound that, on account of its unusual electrical properties, is beginning to be considered for the manufacture of computer chips. The researchers zeroed in on one of the spots that appeared on their scanning micrographs and found that it looked like a snow-covered hill rising out of a plowed field. They interpreted the furrows as orderly rows of gallium alternating with arsenic atoms, and the hill as a lone oxygen atom clinging to the gallium arsenide. When the polarity of the voltage between surface and STM probe was reversed, so that the direction of the tunneling electron flow also reversed, the picture was inverted too: The hill became a crater. This problem is actually an old one, dating at least as far back as 1664. In *Micrographia,* the famous handbook on microscopy, the English physicist Robert Hooke remarked that "it is exceedingly difficult in some objects to distinguish between a prominency and a depression, between a shadow and a black stain, or a reflection and a whiteness in color."

In the case of gallium arsenide the problem is electrical. The imperfection on the regular surface is indeed caused by the bonding of an oxygen atom, the first step in the oxidation process, but the effect of the contaminating atom is electrical rather than structural. The oxygen atom doesn't cause much of a hill or a valley in the surface it attacks, but its electrons do create a large pool of negative charge, which the STM registers as either a hill or a valley, depending on which way the electrons are flowing. This solution to the puzzle was supported by theoretical calculations, as well as examinations of the surface by more conventional means, and served as warning that the STM commingles the structural and electrical properties of the atoms and molecules it depicts.

In the image of benzene on a rhodium surface the collection of white, lumpy mounds that appear in the micrograph undoubtedly corresponds to a layer of benzene. But what about the curious smudges in the middle of each mound that look like the holes in doughnuts? Are they actual indentations or merely the accumulation of excess negative charge created by the forty-two electrons that surround each molecule of benzene?

Ambiguities in the interpretation of images are not unique to the STM. In fact, they are the rule, rather than the exception in science. Whether the object is a virus seen through an electron microscope, a distant galaxy explored by radio telescope, or a fetus observed in the womb by means of ultrasound, each observation must be translated into a coherent image with care. From among the many competing interpretations of the same picture, scientists are often forced to choose without any certainty that their pick is correct. Only in the rare cases in which the shape and color and texture of the image can be verified directly, through seeing and touching, can we be sure that object and image agree; but those cases are growing increasingly peripheral to the aims of modern science.

Inasmuch as both the images revealed by instruments such as the STM and the images formed by the human mind on the basis of nonvisual information are indirect and ambiguous, the distinction between imaging and imagining becomes blurred. Both are ways of producing pictures, and both depend on a body of background information, and on hidden assumptions and theories. Thus the difference between Kekulé's dream and the scanning tunneling micrograph is not so much qualitative as quantitative: The micrograph, unlike the dream, reveals the actual dimensions of molecules, measured in billionths of a meter.

In the end the knowledge accumulated from over a hundred and twenty years of research on the chemistry of benzene gives us the confidence to interpret the dark smudges as holes and the pictures of the molecules themselves as rings. Because the structural model of benzene is far better understood than the micrograph, it is Kekulé's dream that confirms the mechanical image, not the other way around.

No one is more aware of the inconclusiveness of STM imaging than its inventors, nor more anxious to improve the diagnostic tool chest. In 1985, a few years before the benzene image was created at the IBM laboratory in San Jose, Gerd Binnig happened to be visiting there from Zurich. The trip gave him time off from his work on the STM and a chance to think about new ideas. Once, while lying on the floor of the house he was staying in, he noticed the subtle surface structure of the ceiling and wondered how he could record it with his STM, seeing that it was made of plaster and wouldn't conduct electricity. He began to consider other ways besides tunneling currents to explore surfaces and came up with a strikingly simple idea: Why not just feel it out gently, the way a blind person's finger traces the outlines of a face? Why not just touch it and record the push with which it resists the touch? Out of such musings the atomic-force microscope was born.

Developed by Binnig together with a team of American colleagues, the new instrument makes use of a sharp diamond tip mounted on a cantilever arm, like a phonograph needle. The purpose of the cantilever is to reduce the force with which the tip presses against the surface. If that pressure amounted to more than the lightest touch, it would tear up the surface. To the surprise of the inventors, however, it turned out to be very easy to make a cantilever gentle enough for the purpose. An arm made of a sliver of kitchen aluminum foil, fixed at one end, can be bent at the other end by the unimaginably feeble force that operates between individual atoms. The only problem would be to detect the almost imperceptible movement of the diamond tip as it slid over bumps as small as single atoms.

Of course Gerd Binnig was the right man to overcome that particular hurdle. Had he not helped to invent an instrument that is specifically designed to measure displacements of atomic dimensions? Accordingly the original atomic-force microscope incorporated an STM to record the deflection of the metallic cantilever that held the diamond probe. The new microscope substituted actual contact for an electrical current and thereby reduced the ambiguity between electrical and structural features inherent in the images created by the STM.

And yet, what exactly does actual contact mean? When our fingers encounter someone's skin, or when two billiard balls collide, the concept is clear enough. But think about what happens when you pat your dog. At what moment do you touch it? When your hand first encounters the tip of the farthest stray hair that happens to be sticking up; when you touch a dozen hairs; when you feel hair covering your palm; when you first become conscious of the resistance that the fur offers to the pressure of your hand; or when you begin to perceive the body below it? Or is it rather when the dog first senses your hand on its back? Any one of these definitions is acceptable, which implies that none is entirely satisfactory.

At the atomic level even the intuitively appealing concept of touch turns out to be anything but simple. In fact, nothing really ever touches anything else. When billiard balls approach each other, the distance between their outermost atoms decreases, and in the same measure the repulsive force between them increases and slows them down. Eventually they stop momentarily, and then bounce apart and recede from each other. In the scientific way of thinking the focus is removed from the idea of touching and is vested instead in the way in which motion is arrested, be it in the approach of billiard balls or in the lowering of your hand onto Fido's back.

With this interpretation of touching, the question becomes, What forces stop the diamond tip of an atomic-force microscope as it approaches the surface under it? The answer is that there are a number of them, and all but one are electrical or magnetic in origin. Chemical forces, which is to say the forces between atoms and molecules, can be understood almost entirely in terms of electrical attraction and repulsion, and magnetism. The combination of these fundamental effects into actual interatomic forces can be exceedingly complicated, which is what keeps theoretical chemists in business. But in principle microscopic atomic forces are the same as the familiar forces that hold magnets on refrigerator doors and make your hair stand on end when you brush it on a dry winter's day.

The sole exception, the atomic force that has no counterpart in everyday life and cannot be understood in terms of electricity or magnetism, is quantum mechanical in origin and remains utterly strange. The exclusion principle decrees that no two electrons can be in the same state of motion—that is, they can't have the same position and speed—at the same time. The exclusion principle directs the structure of atoms like a traffic cop: It keeps electrons from crashing into one another.

Externally, in the interaction between entire atoms, the exclusion principle acts as a kind of repulsive force that prevents the two electron clouds from intermingling; they never touch. In short when you pat a dog, when billiard balls collide, and when the tip of a needle encounters a surface, the forces that come into play form a Gordian knot of classical and quantum mechanical effects, and it is impossible to determine exactly which one of several influences finally prevents one object from overlapping another. Consequently, while the atomic-force microscope succeeds in escaping the gross ambiguity between structural and electrical properties that afflicts the STM, it suffers from even more subtle problems of interpretation.

How, then, does a physicist imagine an atom? The process might be compared to the steps involved in building a friendship. Suppose the acquaintance begins through correspondence: letters can reveal a lot about a person and can help create a detailed mental picture. But if a photo is included in the letter, one sees a completely new aspect of the person. Or if there is a phone conversation, the sense of hearing begins to contribute its part to the developing image of the friend. When one finally meets this person, the once-sketchy image is fleshed out. Even so, it continues to be revised and improved as the friendship, and with it understanding, evolves. Mental images of friends are not formed instantaneously, but grow slowly, nurtured by all the senses and every conceivable source of information about the person.

In the same way there is not a single definitive image of an atom. The lessons learned from structural chemistry, from quantum mechanics, from the STM and its offspring, the atomic-force microscope, as well as from countless other experiments in atomic physics, all contribute to our rapidly forming image of the atom.

Looking at atoms from the outside, the view is beguiling. At the level of polyatomic molecules and large crystals, we find complex and fascinating structures. Rings, repetitive lattices, orderly arrangements interrupted by unexpected imperfections, helices and other geometric figures of every imaginable design abound. But some of the fine detail is missing. Individual atoms can be distinguished in these structures, but they invariably have the nondescript look of furniture draped in painters' tarpaulins. The temptation to climb into the picture, to tear away those covers, or at least to lift their corners, in order to expose the real atoms underneath, is almost palpable. One hopes fervently that microscopy will continue its downward progress from the molecular to the atomic level, and from there to the subatomic realm, so that future generations may enjoy the privilege that has been denied us so far.

In contrast to the frustrating lack of definition of the view from the outside, the perspective from inside the atom is astonishingly detailed, even though it is impossible to imagine a mental picture of it. Quantum mechanical calculations, refined by bigger and faster computers, are furnishing ever more reliable wave functions of the atomic electron clouds. In principle these contain all the information one can possibly obtain about the atom, but when it comes to intuitively comprehensible representations of the atom, we haven't learned to ask the right questions yet. We know everything about the structure of the atom except its meaning. As diagnostic probes such as STMs begin to delve beneath the atomic surface, the wave function will become increasingly relevant to the interpretation of the image until the two views—from the outside in and from the inside out—eventually become reconciled in a new atomic icon.

But the success of this dual strategy for capturing the image of the atom is not assured. A visual representation of atomic structure acceptable to common sense may be an impossibility because quantum mechanics has a way of protecting its secrets. A low-voltage STM needle and a gentle atomic-force stylus tickle the surfaces they explore without disrupting them appreciably, but when the voltage is cranked up on the STM, and the cantilever of the force microscope is stiffened, so that the electrons and the diamond tip push deeper into the atomic interior where quantum mechanics reigns, the intruding probes begin to modify the samples they are supposed to survey. What is measured, then, is not an atom, as originally intended, but something entirely different and much more complex: a composite system consisting of an atom and a stylus. The observer begins to intrude upon the observed. Studying an atom is like examining the surface of a mirror. If you scrutinize it too closely, you suddenly realize that you are looking at yourself.

In order to picture the elusive atom, we must contrive to slip through the narrow doors of nature's house, as the English polymath Thomas Harriott put it back in 1606. He managed to achieve the feat by the force of his imagination and beckoned the astronomer Johannes Kepler to follow him; until recently that was the only way, and it worked wonderfully well. Kekulé's dream, inspired by a murder trial witnessed in his youth, produced atomic images of such cogency that they survived to corroborate and confirm the interpretation of the ambiguous micrographs made by modern imaging technology. Today imaging is replacing imagining: The STM and the atomic-force microscope provide tantalizing glimpses of the atomic landscape. Our eagerness to explore that world recalls an incident in *Alice's Adventures in Wonderland:*

> Alice opened the door and found that it led into a small passage, not much larger than a rat-hole: she knelt down and looked along the passage into the loveliest garden you ever saw. How she longed to get out of that dark hall, and wander about among those beds of bright flowers and those cool fountains, but she could not even get her head through the doorway.

5

The Atomic Landscape

What Alice longed for, Dorothy found unexpectedly. The scene, as recorded in the 1939 movie *The Wizard of Oz,* is one of the most magical moments in cinema. As the frightened Dorothy opens the door of her wrecked cottage and steps into the wonderland of Oz, the film changes abruptly from black and white to color. The amazement on Dorothy's face as she enters that exotic world of brilliant blossoms and sparkling-blue ponds mirrors the enchantment of the viewer. Part of the impact of the dramatic switch derives from its ironic reversal of perceptions, an effect that must have been even more overpowering when the film was released than it is today. The so-called realistic scenes in Kansas shot in black and white yield to the technicolor fantasy of Oz, whereas by intuition it should be the other way around. The real world, no matter how bleak it may be, is endowed with color, while the artificial dream world of photography and cinema fifty years ago was predominantly black and white. By depicting the real in unreal shades of gray, and the fairy-tale land in true color, *The Wizard of Oz* transports us into a world we never believed we could enter. When color pictures of atoms became available in the late 1980s, they had a similar effect.

Color is a powerful seducer. Beginning with the prehistoric cave painters, artists have used color both to convey visual information and to evoke emotional responses. They know, consciously or subconsciously, that color speaks directly to the soul in a way that pure shape and form do not. Even scientific subjects that might otherwise be regarded as dry and forbidding can acquire a strong emotional appeal by being endowed with color, as modern images of the solar system eloquently testify.

A decade before the advent of artificial satellites the British astronomer Sir Fred Hoyle predicted, "Once a photograph of the Earth, taken from outside, is available—once the sheer isolation of the Earth becomes plain, a new idea as powerful as any in history will be let loose." Twenty years later lunar orbiters sent back such pictures, but none had the impact on the popular imagination that was engendered by the first color picture of the planet, the famous blue marble floating in the blackness of outer space.

Taken in November 1967 by a stationary satellite over Brazil, this picture remains the most popular in NASA's huge collection. Hoyle, trained to extract profound insights from inconspicuous and often grainy black-and-white photographs of heavenly bodies, could not have foreseen the manner in which color would bring the picture home. So vivid was the photograph, with its bright blue oceans, tan land masses, and wispy white clouds surrounded by black space, that its effect was almost the opposite of what Hoyle had predicted. Instead of emphasizing our isolation, the image gave us an unexpected feeling of togetherness and interdependence, a message conveyed largely by its engaging colors.

A dozen years later, when Viking 2 reached the surface of Mars and began to return pictures of the surrounding landscape, it was color again that captured the popular imagination. These images showed rocks and soil of an almost unbelievable reddish hue, a caricature, it seemed, of the red planet's surface. But then NASA revealed the simple and direct way in which it had calibrated the colors sent back in digital code. When the lander was built, a palette of standard colors, measuring about a foot long by six inches wide, had been painted on one of its legs. After the craft had landed on Mars, the on-board TV camera swung over to look down at it, and its electronic signals were adjusted until the image matched an identical palette in a laboratory back on

Earth. Thus reassured of the reliability of the color reproduction, we understood at last that the red was true and that Mars derives its martial countenance not from human blood, as ancient mythology implied, but from the iron oxides in its soil.

Add to the compelling images of Earth and Mars some of the photos of the outer planets and their satellites sent back by the two Voyager probes: Jupiter's swirling eddies of purple around the great mysterious "red spot," the ghostly green tinge of the majestic rings of Saturn, and the pizza-pie surface of Jupiter's moon Io adorned with the sulfurous plumes of active volcanoes. The fuzzy blue dot of Earth seen from the edge of the solar system, an insignificant atom in the vastness of space, finally confirmed Hoyle's prediction.

Because color is a common and familiar phenomenon, it relates unreachable objects to the familiar world around us and makes them accessible. Without color the pictures of the solar system would remain cold artifacts of high technology, as unappealing as reams of computer output; color makes them come alive. The distant planets and their moons may be strange, but now that we've seen them in color, they have become real. The same transformation is happening to atoms, with the fundamental difference that the colors of atoms are not natural but artificial.

As an illustration of this crucial distinction consider the contrasting uses of color in astronomy and mathematics. The colors of astronomical images may be influenced by the optical properties of the cameras that make them and by the complexities of electronic transmission back to Earth, but in principle there is no arbitrariness in them. They are fixed by nature, and it is the business of imaging specialists to devise techniques of calibration to eliminate unwanted chromatic distortions. In mathematics, on the other hand, everything is artificial, nothing real. Nevertheless the power of color to appeal to human feelings endures even in this austere discipline.

An impressive example is the portfolio of pictures in the center of James Gleick's popular book *Chaos,* which depicts a graphical representation of a numerical relationship called the Mandelbrot set. Just as the description of the vagaries of the stock market become more vivid when they are displayed graphically, the Mandelbrot set, a complex collection of special points in the plane that crops up in chaos theory, becomes more understandable when it is pictured in color. Its images are like nothing seen before in art or nature, combining vigorous patterns with the most subtle variation of detail down to the limitations set by the weakness of the human eye.

Remarkably the Mandelbrot set itself is a strictly numerical concept in which color has no relevance. The computer was instructed to associate specific colors with certain intermediate numerical values used in the creation of each pattern, and to display them on the screen. Each color represents nothing more than the speed with which the calculation converges to its final answer for a particular spot in the picture. While the shapes of the images are dictated by the mathematical formula that defines the Mandelbrot set, the choice of colors was made with only aesthetic considerations in mind. Since they have no mathematical or physical significance, they are called false colors and in this case happen to echo the same principles that my wife urges me to observe in the selection of ties: Pink and gray go together, and so do blue and green, but don't mix red with pink. The results are as appealing to the senses as they are intriguing to the mathematical mind.

Even though false colors lack the meaning of true colors, they can be an aid to understanding and to discovery. By guiding the eye, they can reveal hitherto hidden patterns, and by flagging structural features, they can help in tracking them through complex sequences of pictures. Just as brain surgeons find false colors helpful in finding the boundary between healthy and diseased tissue in CAT scans, physicists are beginning to recognize the value of color-coding in their work.

With false color demonstrated in other applications, it was a natural step to display the computer-generated STM images in color as well. The atomic landscape, whose outlines were first faintly suggested a decade ago by the squiggles of the black-pen recorder of Binnig and Rohrer's first STM, can now be seen in the same brilliant colors as the images produced in other contexts.

The use of color in atomic images, false though it is, has a profound impact: It restores to atoms an element of reality that they had lost. Besides shape, texture, weight, and hardness, real objects have color, no matter how drab it may be. The atoms of Democritus can be imagined as being endowed with color, but the modern atom has none because one of the functions of atomic models is to explain color in mechanical terms. Isaac Newton pointed the way when he speculated about the particles on which "the colors of natural bodies depend." Later, when Bohr invented a detailed model of the atom, the production of colored light was associated with quantum jumps of electrons to lower orbits, and quiescent atoms became colorless specters. Thus atoms were robbed of one the most obvious sensual qualities of real objects—color. The mind struggles to imagine the Bohr atom in pictures devoid of color, so it reverts to the convention of print and renders it in black and white, like the opening scenes of *The Wizard of Oz*. The return of color to the atomic landscape invests it with a false aura of reality.

The scientists who produce STM images color them for reasons other than emotional appeal or the hope of making them seem more real. A simple purpose of color coding is the identification of different species of atoms. When location or other characteristics allow unambiguous distinctions to be made, each atom has its own color. Thus, for example, a 1987 micrograph of a surface of the high-tech compound gallium arsenide pictures arsenic in red and gallium in green. Theoretical calculations had predicted the arrangement of the atoms in alternating rows (just as Kekulé predicted the appearance of benzene molecules), and the micrograph confirmed the prediction. The color scheme adds visual contrast and renders the picture memorable. It looks like a photo of a stack of red-and-green Christmas-tree balls neatly packed away for the summer. Another picture, made a year later, shows a long yellow DNA molecule snaking its way like some revolting worm across a sickly-green silicon substrate. (If the colors were more harmonious, the molecule might evoke a golden necklace on a velvet cushion instead.)

A more instructive and less flamboyant use of false color is borrowed from cartography. Modern maps show altitude and ocean depth by means of delicate variations in tint, rather than by the black contour lines and hatched shadings of an earlier era. The same technique can be applied to representations of the atomic landscape. Indeed, in the gallium arsenide picture this was done for each atomic species separately. Each single atom of gallium or arsenide grows brighter toward the top, as though a light were shining across the stack and giving it an uncannily three-dimensional look.

There is a third scheme for coloring STM images that falls, in a curious way, somewhere between the concepts of true and false colors. The notion of color, when the image is made by electrons, can arise indirectly, by analogy to the color in photographs made with ordinary light. In physics color is quantified by three closely related concepts: wavelength, frequency, and energy. Starting at the red end of the spectrum and proceeding toward violet, wavelength decreases, frequency increases, and energy, which is related to frequency by Planck's law, also grows. Thus each photon can be characterized interchangeably by its color, wavelength, frequency, or energy—the attributes that together define the physiological, optical, and mechanical properties of light.

In the STM the energy of tunneling electrons is determined by the voltage applied between the probe and the sample: The higher the voltage, the more energetic the electrons. Guided by the fact that electrons are waves, too, physicists have begun to borrow the terminology of light and refer to a new concept that might be called electro-color, even though electron waves are not visible to the human eye. Images produced with different voltage settings differ in detail, both because higher energy allows electrons to penetrate more deeply into the structure of the sample and because the distribution of electrical charges in the sample affects electrons with different energies in different ways. So STM images made with different voltages correspond to pictures made under the illumination of different electro-colors. When the computer is programmed to assign a different false color to each voltage setting, or electro-color, the analogy between light and electrons is exploited to the fullest extent. After each image is created under a certain illumination determined by the voltage setting, and displayed in one false color, several images can be combined into a multicolor picture, like a composite of photographs taken with different color filters. In this way atomic landscape is captured not by color photography but by electro-color micrography.

Whatever the meaning of the electro-color concept, or of the false color selected by the computer programmer, the images of the atomic landscape are wondrous to behold. In beauty and evocative power they rival the pictures of the land of Oz, the photographs of the solar system, and the synthetic illustrations of chaos. But while those depict, respectively, imaginary, inaccessible, and abstract scenes, the atomic landscape is real and in the literal sense as close as our fingertips. It is a territory that has been intellectually conjectured for millennia but forbidden to the senses, until now. Like Dorothy in the land of Oz, we look around in wonder.

Here a complicated organic molecule rises out of a plane of graphite, like Mount Everest out of a snowy plain, its valleys in shadow and its peaks gleaming in brilliant computer-generated sunlight. There a row of molecules looks like a phalanx of blue ocean waves capped by yellow-and-white foam, while the troughs between the crests seem filled with pools of sparkling claret. A micrograph of the surface of silicon appears at first glance to show a heap of blueberries, until the eye picks out the repetition of most unblueberrylike hexagonal ringlets of six atoms each. A picture of clumps of gold atoms deposited on a surface is colored, appropriately enough, golden yellow and orange. If it weren't for the substrate, which is blue, the picture could be mistaken for a close-up of the surface of a German crumb cake.

Some pictures are less homely and more enigmatic. A group of purple iodine atoms is interconnected by a geometrically perfect lattice of dark blue arms that reach out, six per atom to six neighbors. Kekulé might have recognized the pattern from his first dream. But one of the iodine atoms is missing, and the hole where it should have been signals an accident during the formation of the crystal or a collision with some atomic projectile. The color scheme calls for green at the outermost bulges of the surface and then yellow, followed by orange at the deepest points. Consequently the hole glows with a yellow light that issues from some subterranean orange fire, rendering the scene vaguely surrealistic. A more elegant display is provided by six xenon atoms nonchalantly leaning against a step on a platinum surface. They, along with the stepped surface, appear to be shrouded under a smooth silk sheet, the color of platinum, that hides the atomic interior from prying eyes. Less sleek and more woolly is a picture of a layer of liquid-crystal molecules, of the kind used in laptop computer displays, in which two kinds of fuzzy shapes alternate in rows of garish red and green. Shocking though it is to the eye, the image looks soft and inviting to the touch.

And why not touch it? If technology can enlarge the atomic landscape sufficiently to bring it to our feeble eyes, why not enhance its features until we can feel them too? That question was asked by a team of researchers at IBM's Thomas J. Watson Research Center in Yorktown Heights, New York, where many improvements of the STM have been developed.

The task of building a machine that enables humans to feel atoms is imposing. A large molecule is about a million times smaller than a human finger. Although a million doesn't sound like much in this age of gigantism, it helps to imagine a probe that is a million times *larger* than a finger: It would measure about five miles wide by twenty-five miles long—a formidable digit. The IBM scientists succeeded in going in the opposite direction.

They call their device the magic wrist. It consists of two electrically connected instruments—a robotic manipulator and a conventional STM. When the operator places his fingers on the manipulator, which looks like a hexagonal aluminum box the size of a teapot and is mounted on a round baseplate, he can guide it horizontally through distances of a few millimeters. This motion, reduced electronically by a factor of a million, is transmitted to the STM probe, which slides across a sample surface in response. At the same time the vertical motion of the tip, as it encounters the atomic hills and valleys of the surface, is exaggerated by a factor of a million and fed back into the manipulator. As a result the operator can feel the surface roughness of the atomic landscape with his fingertips.

The technological promise of the magic wrist is boundless. It will be made into an instrument that can move and manipulate individual atoms, perform surgical operations of a hitherto unimaginable delicacy, assemble designer materials for all manner of purposes, and build the smallest conceivable computers and other machines. At present the magic wrist is in its infancy and enables its operator to feel nothing smaller than clumps of several atoms sprayed onto a smooth background surface, but the philosophical and psychological significance of the device is as immense as its future usefulness.

The sense of touch afforded by the magic wrist provides the ultimate confirmation of reality. James Boswell, the biographer of Dr. Samuel Johnson, gave a classic example of such a confirmation: "We stood talking for some time together of Bishop Berkeley's ingenious sophistry to prove the non-existence of matter, and that everything in the universe is merely ideal [composed of ideas]. I observed, that though we are satisfied his doctrine is not true, it is impossible to refute it. I shall never forget the alacrity with which Johnson answered, striking his foot with mighty force against a large stone, till he rebounded from it, 'I refute it thus.' "

Actually the bishop had already gone a step beyond the good doctor's vivid demonstration of what it means to be real. In his *Essay Towards a New Theory of Vision,* he had distinguished between objects that can be "realized" through touch and others that can be detected by eye alone, including those that are "perceived by the help of a fine microscope." To the former he grudgingly afforded a certain measure of reality, but the latter he dismissed as immaterial, fit only for "the empty amusement of seeing." Atoms made visible by scanning tunneling microscopes would therefore have escaped Dr. Johnson's vigorous defense. He could not have kicked them, and Berkeley would have won his point that they are mere theories. But with the magic wrist we can flick a finger at an atom of gold and triumphantly reply to Berkeley and the old skeptic Ernst Mach, as well as all who have ever doubted the atomic doctrine, "Atoms are real, and I demonstrate it thus."

For scientists atoms have been as real as Dr. Johnson's stone for a long time. The new accessibility of the atomic landscape brought about by color images and the touch of the magic wrist will make it real for nonscientists as well. We will witness the expansion of our collective consciousness down to a scale that had for ages been the exclusive preserve of philosophers and professional physicists.

Theories and formulas are sufficient for minds trained in abstract thinking, but ordinary people understand by seeing and feeling. One of the greatest of abstract thinkers, the philosopher Plato, expressed this forcefully in the *Timaeus,* which ironically had inspired the young Heisenberg, who later shunned models. Commenting on the motions of the planets, Plato wrote, "To describe the evolutions in the dances of these gods, their juxtapositions and their advances, to tell which one came into line and which in opposition, to describe all this without visual models would be labor spent in vain."

The universal appeal of model airplanes, trains, and ships, no less than the popularity of modern planetariums, prove Plato's point. Visual models render complex systems seemingly simple and foster more complete understanding. For this reason images of the atomic landscape will eventually play an important role in teaching, and may even replace Bohr's atom as popular icons.

The way scientific models capture the imagination of nonscientists has always been of interest to me, so I responded with delight when a friend told me about the atomic model of the sculptor Ken Snelson and offered to take me to his studio in New York to introduce us.

Snelson is best known for his tube-and-wire constructions, which stand in museums and public places throughout the world.

Made entirely of straight sections of stainless-steel tubing of various lengths, held together by stretched aluminum wires, they look at first glance like ordinary scaffolds. But on second glance a miracle appears: The tubes don't touch each other. They stand up and dance into the sky, enveloped in a wispy spiderweb of wires, in defiance of the laws of gravity. They are Indian rope tricks executed with high-tech materials, magical, powerful, and entrancing.

I met Snelson in his loft in Manhattan. Amid a jumble of machine tools, bookshelves, an enormous computer console in a corner, and miniature models of his monumental sculptures covering every available open spot, we sat at his work table and drank tea. He is short and powerfully built, with a workman's rough hands and an open, square face topped by a thick thatch of dark hair that belies his sixty-odd years. Snelson told me that what fascinates him most in his work is the way pressure and tension, represented by tubes and wires, play off against each other to create a structure. Structure fascinates him. "I'm not even sure I'm a sculptor," he said, "I'm interested in three-dimensional space. I'm a structuralist."

Gifted with a profoundly analytical mind, he began to think about the physical foundations of the structures he became so adept at building. In this way he arrived at what he calls the structuralist's ultimate question, the problem of the structure of atoms. Having identified the fundamental ingredients of structure in the macroscopic world we live in, he wanted to do the same thing at the microscopic level of atoms. "If I could figure out the structure of atoms," he said with a twinkle in his eyes, "I could build the universe."

For thirty years, while his reputation as sculptor has grown to world renown, Ken Snelson has been thinking about atoms. Not being a scientist, nor even comfortable with mathematics, he could not follow the technical literature on quantum mechanics and its explanation of atomic structure. The popular descriptions of these arcana left him unsatisfied, so he proceeded to invent his own model of the atom, "my fantasy atom," as he calls it. The driving force behind this effort is the artist's passionate desire to *see*, which parallels the scientist's passionate need to *know*. "The mind hungers for pictures of everything," Snelson has written, "atoms, no less than trees, flowers and creatures."

His motivation is precisely the same as that of Democritus, Rutherford, and Heisenberg: the craving to understand matter at its most fundamental and learn what the world is made of. Only the methods of inquiry differ. Where Democritus applied philosophy, Rutherford the experimental approach, and Heisenberg the tools of mathematical analysis, Snelson uses art. The symbols of the first three are words, numbers, and equations, Snelson's are visual forms. His creations presage a time in the future when atoms will preoccupy poets, artists, and the world at large as much as scientists.

On a shelf in the back of the studio Snelson kept his primitive models which he made from wood, metal, glass, plastic, and paper, before he discovered computer graphics in 1985. The flexibility of this new medium allowed him to become the first major artist of the atomic landscape.

Snelson's creations differ from STM images primarily in their depiction of individual atoms. In the actual micrographs they always seem to be covered up, like Victorian bathers, so no part of their underlying anatomy is ever revealed. Snelson's atoms, on the other hand, do not suffer from such modesty. Since, as an artist, he is free to invent, he has created a scheme in which atoms look like translucent spheres with colorful rings representing electrons stuck to their surfaces. Snelson's atoms are vaguely reminiscent, except in architectural detail, of some of the older models of the atom that were popular with physicists before the advent of quantum mechanics. But they are beautiful despite their lack of scientific justification, and satisfy Snelson's need to express atomic structure in visual terms.

In 1989 Snelson created a piece called "Kekulé's Dream," which displays the structure of the benzene ring according to the rules of his atomic scheme. The molecule looks mechanical and austere, a little like a complicated chain of steel rings of the kind magicians use on stage, except that they are not interlinked. To me the picture represented a synthesis of Kekulé's vision and the computer-generated micrograph made by the STM. But Snelson is an artist with a wide-ranging and fertile imagination, and this one work didn't exhaust the possibilities inherent in the idea of the benzene ring.

The last time I visited him, in the fall of 1990, he took me straight over to his computer console to show me some work in progress. He switched on the oversized monitor, and, after a hurried search through menus and directories, the screen exploded into a monstrous mass of writhing green creatures. They were snakes—sleek, powerful, evil-looking serpents—shiny and scaly, like medieval manuscript illuminations, and I was astonished to see that they were biting their own tails.

Thus Snelson had returned inexorably to the roots of the human imagination where science and art meet. After a foray into the frontiers of the atomic landscape he had come back to the safe haven of the subconscious. But somehow this mythological image was closer to the actual benzene micrograph taken in California than his earlier model. Coiled serpents capture the essence of the molecule symbolically, but they also suggest an attribute of real molecules that is missing from mechanical models—the element of mystery. They remind us that just beneath the surface of the dazzling atomic landscape recorded by modern technology, the paradoxes of quantum mechanics lurk like venomous snakes.

6

Atoms in Isolation

On the surface the atomic landscape looks uncannily familiar. Atoms and molecules are so far removed from daily experience, and their quantum mechanical descriptions so foreign to ordinary language, that when we finally get to see them, we expect to find an otherworldly and astonishing scene, an enigma like a platypus. But instead we find a tray of lumpy doughnuts, a stack of red-and-green Christmas ornaments, a revolting worm, a phalanx of blue ocean waves, a heap of blueberries, a German crumb cake—familiar, commonplace objects. The realm of atoms appears to be just like the everyday world, transposed down to some incomprehensibly tiny scale. The surprise is that it doesn't look surprising—the way the Mandelbrot set does, for example, or the pimpled surface of Jupiter's Io. Atoms look too ordinary.

The quality of continuity is the most surprising feature of those images. Ordinary objects appear to consist of continuous, unbroken matter; atoms should not. The logical difficulties of the idea of continuity, such as the questions of the divisibility of solids and the mixing of liquids, led Leucippus and Democritus to postulate a new and different reality, an atomic reality, behind the facade of continuity. But now, two and a half millennia later, when we have finally reached the level of atoms and expect to see a new world, what do we find? Just a tray of doughnuts and a heap of blueberries?

The apparent continuousness of STM images has two fundamental causes. First there is the problem of resolution. No matter how fine the needle of a scanning probe may be, its tip can be no smaller than an atom. This means, in turn, that the pictures it makes are limited in sharpness: the surface of a phonograph record feels smooth because our fingers are too large to perceive the grooves that our eyes can see. As technology improves, sharper images are sure to come, but there is no such thing as an infinitely pointy needle or a picture with infinite resolution. Only in the make-believe world of mathematics is it possible to resolve the details of objects to ever finer detail, to subdivide ad infinitum. In the domain of the atom there will always come a point when two separate features of an object appear as a single one because the probe is too clumsy to tell them apart. This is the crucial difference between portraits of the Mandelbrot set and of the real world.

The second cause for the apparent continuity we see in STM images is that the atoms on the surface of an object are in fact connected. The bands that seem to tie them together and the sheets that seem to cover them have a simple physical origin. In bulk matter, and on surfaces, neighboring atoms bump and jostle each other, and all the while share electrons. Their electron clouds are so intertwined that it is impossible to distinguish, even in principle, which electron belongs to which atom. To make matters worse, metals and other conductors are suffused with electrons that are free to roam over the entire sample. They wash over the surface like ocean waves, obliterating the structural details of individual atoms.

For now, at least, the continuous surface of the atomic landscape is impenetrable to the STM. The electronic mortar that connects the atomic building blocks of matter hides their intrinsic granularity. With that realization the relevance of the atomic landscape shifts from the province of physics to that of chemistry. Studying the structure of various combinations of atoms is really the business of chemists. How do carbon and hydrogen combine into benzene? What is the shape of a DNA molecule? How are the constituents of gallium arsenide arranged? These are chemical questions for which scanning microscopy can provide answers.

But questions of physics concerning the internal structure of the atom itself require a different approach.

If photographs of dunes do not reveal the nature of sand, perhaps a single grain, held up to the light and examined in isolation from its countless siblings, will. Likewise, if the resolution of the STM is limited by atomic interactions, it might help to isolate and scrutinize the atoms individually. Unfortunately until very recently this was not believed to be feasible.

In fact the difficulty of capturing single atoms was considered to be more of an advantage than a drawback, because it provided a convenient escape from the paradoxes of quantum mechanics. Faced by such conundrums as spread-out wave functions that describe the positions of point particles, physicists assumed that wave functions of individual atoms don't really have a meaning, and made an analogy with statistics. The probability of throwing heads in a large number of coin tosses is a perfectly well-defined, easily measured quantity. In fact it is found to have the numerical value of one half with astonishing accuracy. But applied to a single toss, this probability predicts nothing, explains nothing, and cannot be measured, so it is meaningless.

In the same way, it was argued, the wave function of a single atom, while it can be calculated, lacks concrete meaning: For a single observation of that atom the wave function predicts nothing. The only meaningful comparison between theory and experiment is a probabilistic one that entails the observation of a very large number of atoms, either all at the same time, as in the case of light shining through a beaker full of liquid, or in succession, as in Young's double-slit experiment with a feeble source.

The most ardent spokesman for this point of view was Erwin Schrödinger, scientifically the most conservative of the architects of the original edifice of quantum mechanics. In 1952, when the theory was already fully mature and universally accepted, he published a paper in the *British Journal for the Philosophy of Science* under the title "Are There Quantum Jumps?" Using italics for emphasis, he wrote "we *never* experiment with just *one* electron or atom or (small) molecule. In thought-experiments we sometimes assume that we do; this invariably entails ridiculous consequences." To explain the scintillations of individual alpha particles observed in the spinthariscope, and the distinct tracks of single electrons, nuclei, and other more exotic particles recorded by photographic emulsions, he wrote, "It is fair to state that we are not *experimenting* with single particles, any more than we can raise Ichthyosauria in the zoo. We are scrutinising records of events long after they have happened. . . . We can never reproduce the same single-particle-event under planned varied conditions; and this is [normally] the typical procedure of the experimenter."

In the days when appeal to authority constituted sufficient proof of philosophical propositions, this categorical statement by so eminent an expert, in so venerable a publication, complete with those impressive italics, might have settled the matter. In the context of modern science, however, Schrödinger's claim was more likely to serve as a provocation than an edict. Just four years after the publication of his paper, Hans Dehmelt, a young German physicist just hired by the University of Washington in Seattle, took up the challenge and began a quest for isolated particles that would earn him the Nobel Prize thirty-three years later.

Perseverance is the defining trait of Dehmelt's character. Today he is a gnomelike little man, with a round, bald head, tiny eyes set in a wrinkled face, and two sharp vertical creases between his eyebrows that confer upon him a permanently quizzical expression. As if to demonstrate his independence from the vulgar world, he sports bushy white sideburns all the way down to the corners of his mouth, which give him the appearance of a character out of Dickens. But there is nothing Pickwickian about Hans Dehmelt: Behind that kindly, naïve smile hides a razor-sharp mind and a ferocious drive.

Dehmelt's entire scientific career, which has been going strong for nearly fifty years, has been inspired and guided by an idea of stunning simplicity. He refers to this concept as "a single atomic particle forever floating at rest in free space" and clearly recalls its origin. In the 1940s, when he was a student at the

University of Göttingen, his teacher, Professor Richard Becker, drew a dot on the blackboard, and declared, "Here is an electron." Dehmelt remembered Werner Heisenberg's admonition that physics should have no traffic with unverifiable abstractions and should stick, insofar as possible, to experimentally measurable, or at least observable, quantities. Heisenberg had borrowed this philosophical principle from Einstein's theory of relativity, and it had stood him in good stead. For Dehmelt—and on this point Schrödinger would also have agreed—the lone dot on the blackboard represented an abstract fiction that had never been realized experimentally. But rather than trying, like Schrödinger, to bury it under an avalanche of statistics, Dehmelt decided to do the opposite. He set out to produce that electron.

The phrase "a single atomic particle forever floating at rest in free space" was chosen with care. It refers to a microscopic particle, such as an electron, an atom, a nucleus, or in fact to any one of the hundreds of denizens of the subatomic zoo. If the same phrase were applied to macroscopic objects, it would represent the fundamental starting point of mechanics, realized, to some degree of approximation, by a marble on a frictionless surface or a star in outer space. The terms *single* and *at rest* are used to distinguish Dehmelt's idea from techniques involving aggregates of atoms, such as scanning microscopy, and those in which streams of particles, such as the jet of electrons in J. J. Thomson's vacuum tube and Ernest Rutherford's beam of alpha particles, are used. *In free space,* finally, serves notice that the particle would not be disturbed by collisions with stray gas molecules, by the jostling of neighboring particles, or even by outside influences such as Earth's magnetic field. Unless the experiment is conducted in outer space, however, Earth's gravitational field is, alas, inescapable, but fortunately in most circumstances irrelevant.

Dehmelt's image is a powerful universal conception. It is what every physicist envisions at the mention of the word *atom,* and both Heisenberg's and Schrödinger's quantum theories refer to this object. It is also what the prophets of atomism—Leucippus, Democritus, Lucretius, Harriot, Newton, Bernoulli, Dalton, and all the others—had in mind. It is the dot thoughtlessly flicked on the blackboard every day in every physics laboratory and classroom in the world. Surely, thought Dehmelt, this object must really exist.

The first particle upon which Dehmelt focused his attention was the electron, and he knew the task of capturing one would not be easy. The electron is an elusive little speck, weighing a mere 10^{-30}, or one nonillionth, of a kilogram. Its charge is so minute that 10^{19}, or ten billion billion, electrons are required to power a hundred-watt light bulb for just one second. As for size, the electron has none at all: It is a mathematical point. These attributes defy the human imagination and take some getting used to before they become the solid truths upon which to build further understanding. How helpful it would be, then, if one of them could be tamed and given a place in our everyday world, thus assuring us that they exist.

In 1973, seventeen years after he first conceived the idea, Dehmelt succeeded. The paper in which the capture of a single electron was announced bore the names of Dehmelt and two assistants at the University of Washington: Philip Ekstrom and David Wineland. David, my host in Boulder, was one of the pioneers of particle trapping. When I asked him to tell me about his former boss, he became evasive. Dehmelt's ego, I gathered, is as legendary as his persistence and his sideburns.

The device in which Dehmelt caught an electron was first developed in 1936 by the Dutch physicist Frans Michel Penning as a means of confining electrical currents for radio tubes and was later modified into an exquisitely sensitive apparatus for the manipulation of individual particles. The Penning trap is a little box of copper and glass, smaller than a light bulb, that confines an electron in a vacuum between two negatively charged plates. A magnetic field surrounding the plates deflects the electron's path, thereby preventing the particle from escaping sideways and hitting a wall, where it would be irretrievably lost. (The magnet is stronger than the one J. J. Thomson used long ago, so the electron's path curls right around and closes on itself.) Repelled upward by one plate, then downward by the other, the electron circles endlessly within the box, bobbing like a horse on a merry-go-round.

Getting a single electron into a Penning trap in the first place was no mean feat. Near the center of the box Dehmelt and his team had mounted a negatively charged metal spike—a miniature lightning rod that sprayed electrons like a fountain. They couldn't see the trapped particles, but were able to monitor their motion by recording the radio waves they emitted. (Whenever electrons oscillate, whether they happen to be in the transmitting antenna of a radio station or in a Penning trap, they radiate.) By careful adjustment of the dials controlling the charge of the plates and the strength of the surrounding magnetic field, the physicists permitted the electrons to escape, one by one, until only one remained. It took a deft touch and a lot of concentration, as in the operation of a pinball machine, but in the end, there it was, a single isolated electron to be watched and pondered.

One of the trapped electrons inhabited the container for ten months before it accidentally collided with a wall and drowned in a sea of copper atoms. By singling it out and observing it patiently for almost a year, Dehmelt's group had succeeded in taming it. The fox told Antoine de Saint-Exupéry's Little Prince, "To you, I am nothing more than a fox like a hundred thousand other foxes. But if you tame me . . . I shall be unique in all the world." Of course there was nothing objectively unique about the Seattle electron, but then it is precisely that sameness that makes taming such a useful process. The feeling of familiarity that the Seattle researchers developed for their guest electron helped them to understand the behavior of all electrons. That they really did feel the emotional attachment that resulted from taming the particles they caught is demonstrated by the fact that they gave some of them nicknames. A positron, positive sibling of the electron, that remained in the trap for over three months in 1987 was called Priscilla, and in 1980 the first atom ever to be photographed in color, a blue barium atom, was named Astrid.

The scientific payoffs of Dehmelt's difficult electron capture experiment are impressive. A measurement of the electron's magnetism, for example, achieved an accuracy of four parts in a trillion, a thousand times better than previous estimates, mainly because the disturbing influence of surrounding particles was eliminated. At least as important, however, is the way in which the experiments reassure us that electrons are not merely convenient mathematical constructs but, contrary to the beliefs of Schrödinger and Heisenberg, real, permanent objects that can be singled out and studied, like grains of sand on the beach.

All previous experiments with electrons either involved large numbers of them or yielded only circumstantial evidence. J. J. Thomson's discovery of the electron, for example, was based on his observation of a thin streak of radiation in the vacuum of a glass tube—which is like studying water molecules by watching a stream from a hose. The pictures of single electron tracks in photographic emulsions, on the other hand, were compared by Schrödinger to the fossil remains of Ichtyosaurs.

An electron in a Penning trap is an altogether different proposition. Although it can't actually be seen, the persistence of its signal over the course of many months endows it with the most persuasive aspect of its existence: its permanence. Hans Dehmelt knows that the same electron he left in its glass-and-metal cage the night before will still be suspended there in the morning. With the utter confidence born of visceral understanding he can even contradict the great Schrödinger, who wrote, "Most theoreticians will probably . . . admit that the individual particle is not a well-defined permanent entity of detectable identity or sameness."

Having tamed the electron, Dehmelt and his group immediately set its sights on the atom itself. The apparatus they chose for this purpose differs a little from the Penning trap. Instead of a magnetic field it uses an oscillating electric field to prevent the charged particle from touching the walls. This device was invented by Wolfgang Paul of the University of Bonn, and earned him the Nobel Prize, along with Dehmelt, in 1989. (Forty years earlier the two men had worked side by side in Göttingen, but their paths had diverged when Dehmelt emigrated to the United States. In the interim Paul has become one of the grand old men of postwar German physics, and so genial a boss that in his latest publications he is proud to list both his sons as collaborators.) Dehmelt miniaturized Paul's trap and developed it into the ring-shaped device I saw in Boulder, which fits on the inscription on a penny with room to spare and is about a thousand times smaller than the Penning trap that first confined electrons.

Technically the proposal to catch atoms in such a trap was so bold that, as Dehmelt relates the story, the officials of an American funding agency that had financed his research for many years simply couldn't bring themselves to believe his claims, and cut off his support. *Non compos mentis* (not of sound mind) Dehmelt claims they considered him, though it is doubtful whether they actually used that phrase. The University of Heidelberg, on the other hand, offered him a visiting position and resources for experimentation, and it was there that he and three collaborators took the first photograph of a single atom in 1979. A year later the charged barium atom, Astrid, was photographed in Seattle, in color.

Unlike the mercury atom I saw in Boulder, which emitted invisible ultraviolet light, Astrid glowed in visible light, and her color was the real thing. Her picture has been widely reproduced in books and magazines, so she is certainly the most famous model among atoms. I hope Primo Levi, the Italian chemist and writer, had a chance to see her picture before he died. In the last chapter of *The Periodic Table* he relates the turbulent history of a carbon atom's journey from its resting-place in a limestone cliff to a cell in his own brain. However, before the story begins, Levi muses, "Is it right to speak of a 'particular' carbon atom? For the chemist there exist some doubts, because until 1970 he did not have the techniques permitting him to see, or in any event isolate, a single atom; no doubts exist for the narrator, who therefore sets out to narrate." Levi the narrator, like Friedrich Kekulé and Ernest Rutherford before him, had no trouble seeing atoms in his mind's eye. Levi the chemist's doubts would have been allayed by Astrid's picture.

Astrid is tiny. She appears as a pale blue dot in the middle of a vast field of pitch black, like the earth seen in Voyager's last shot. But her image is so small that crude reproductions of the picture miss her altogether. Thus a review article by Dehmelt in the proceedings of a 1987 technical conference in Stockholm is accompanied by a mysterious picture, labeled Astrid, which shows absolutely nothing at all. Even atoms can be coy.

The nature of the image of a trapped atom is completely different from that of an atom under an STM. The latter becomes "visible" when tunneling electrons pass through the outer surface of its shell and register as current, which is then used to create a computer-drawn reproduction of the atom. Astrid, on the other hand, actually absorbs and reradiates blue light in the form of photons, which the human eye can see.

In terms of the old Bohr model the process can be explained as follows: One of the intervals between two of barium's many energy levels corresponds to the energy of a blue photon from a laser. When such a photon hits the atom, it is absorbed, and the electron is promoted to the higher of the two levels, but it doesn't stay there for long. Almost immediately it returns to the lower level, emitting another blue photon, which can take off in any direction. Accordingly, when Astrid was illuminated by a narrow laser beam, she lit up in a glowing sphere. In this fashion Astrid absorbed and reradiated several hundred million blue photons per second, of which some tens of thousands reached the camera. Considering that an electron must jump up, and then down again, for every photon that is absorbed and reemitted, it appears that Astrid was not the calm, serene globe she appeared to be but a seething, dynamic little lady, flinging photons in all directions like a whirling dervish.

Color photography and false-color micrography of atoms differ from each other as much as a snapshot of a face differs from what one feels by tracing that face with one's fingertips. But just as we build up the mental image of a friend from all available clues, we try to combine different views of the atom into a reasonable model. The task is made difficult by the fact that quantum theory, which accounts for all atomic measurements and observations, is not written in everyday language. What we must eventually do is to try to grapple with its counterintuitive propositions.

An important concept of the quantum theory that is not

obvious to the eye has already been used to describe Astrid: the notion of the quantum jump. Of course this term is borrowed from the old Bohr theory, with its planetary model of the atom, and the wave equation has no truck with it. The picturesque language Schrödinger would have used to describe what was happening inside Astrid differs drastically from Bohr's language. Instead of saying that an electron occupies one energy level at a time and hops from one to another, Schrödinger would claim that it sits on both of them simultaneously. This is the central mystery of quantum mechanics all over again, the unexplainable fact that an electron can pass through two holes in a screen, or revolve in two orbits about a nucleus, at the same time.

When the atom is in this schizophrenic state, it is receptive to absorption of a photon with an energy equal to the difference between the two energy levels. After it has absorbed such a photon, it can then emit another, identical one—all the while remaining in its ambivalent condition. Schrödinger saw the scattering of light by an atom not as a fitful mechanical process but as a more harmonious phenomenon, as a resonance—like the resounding of middle C on a piano when a nearby flute plays the same note.

This was the kind of language Schrödinger felt comfortable with; quantum jumps disgusted him. "If all this damned quantum jumping were really to stay," he grumbled, "I should be sorry I ever got involved with quantum theory."

And yet, there are quantum jumps. I saw them with my own eyes in Boulder. As I watched that mercury atom, it suddenly blinked off, refusing for a moment to resonate with the ultraviolet light that continued to illuminate it. David Wineland's explanation was that the electron performed a quantum jump into an unreceptive state, in which it could not absorb or emit photons—so the atom became dark. How can this reasonable scenario be squared with Schrödinger's emphatic denial of quantum jumps?

The answer to this question is that both Schrödinger and Wineland are correct, but that they are talking about different things. As long as an atom minds its own business, unobserved by anyone, its wave function evolves according to the rules of quantum mechanics. In describing such an atom, quantum jumps are not only awkward, they are downright incorrect. The atom is like a ripple created by a pebble thrown into a still pond.

But whenever you make a measurement of the state of the atom, the description changes. Whenever an electron is caught in the act of passing through one particular hole in a screen, or occupying one particular orbit in an atom, Schrödinger's wave equation no longer applies. The act of measurement destroys the wave function. That's as it should be, considering that the wave function is really a measure of probability. Observation, or measurement, destroys the very concept of probability, and with it the atomic wave function.

In spite of the controversy that has surrounded the act of measurement since 1926, nobody understands it. The only point of general agreement on this issue is that the wave equation cannot deal with measurements, and this is the very point that allows quantum jumps to return to the debate. Suppose that you measure the state of a single atom repeatedly, always getting the same result. Then, suddenly, between two successive measurements, there is a spontaneous internal rearrangement, and the next measurement records the change. That, in modern parlance, is a quantum jump. Schrödinger could not have objected to this usage, because in this case the series of constantly repeated measurements render his wonderful wave equation meaningless.

In 1986 Hans Dehmelt and his colleagues at the University of Washington published a paper announcing the observation of quantum jumps in a barium atom. At almost the same time his former associate David Wineland, with his team at the National Institute of Standards and Technology in Boulder, published the results of their observation of the same phenomenon in mercury. Seventy-three years had elapsed since Bohr first devised his theory, and the definition of a quantum jump had to be revised, but here at last another fundamental pillar of theoretical physics had been experimentally observed, another prediction realized. The experience was humbling, yet hopeful: If we are patient, the time may yet arrive when we come to grips with quantum mechanics.

Another step toward understanding quantum jumps was

taken in Germany. The blinking of a barium atom like Astrid is slower than that of mercury: It goes dark only once or twice per minute, rather than several times a second, and the darkness persists each time for many seconds. The fundamental mechanism of blinking is the same in both cases, only the arrangements of the energy levels involved in the quantum jumps differ in detail. At the University of Hamburg Peter Toschek, one of the scientists who had helped Dehmelt to take the first photograph of a barium atom in 1979, also discovered quantum jumps at the same time as Dehmelt and Wineland and figured out a clever way to study them. Whenever his barium atom went dark, he fired a beam of yellow laser light at it, which restored it to its original receptive state by a circuitous route of swift internal rearrangements, and instantly the blue light came back on. In this way he shortened the dark times at will and was even able to eliminate them altogether.

Technically the marvel of these experiments is found in the numbers. A single photon of yellow light, absorbed by the atom, suffices to switch on a flood of hundreds of millions of blue photons per second. Engineers call this process amplification and dream of devising gadgets that make use of amplifications as prodigious as this one, which amounts to a factor of several hundred million. A detector of faint yellow starlight built on this principle, for example, would magnify the intensity of the images it receives by that amount and would revolutionize astronomy. But that's for the future. Intellectually the value of the experiment lies in its confirmation of the quantum mechanical calculations and in its demonstration that quantum jumps are real.

The newfound ability to manipulate the inner workings of an atom, to switch it from one state to another and cause quantum jumps to occur on command, will eventually have a powerful impact on all of science. One obvious way in which this will happen is through the measurement of time.

When I visited David Wineland at the Time and Frequency Division of the National Institute of Standards and Technology, he told me that the ultimate justification for his work is the search for the perfect clock. In his terminology a clock is any device that repeats its motion with a steady beat, and a trapped atom is such a device. In the past, time has been kept by the human pulse, the dripping of water from a faucet, the swing of a pendulum, the rotation and revolution of the earth, the motion of our moon and the moons of Jupiter, the vibration of crystals such as quartz, and by countless other repetitive phenomena. Trapped atoms offer the possibility of being used as clocks of unsurpassed accuracy.

Currently the best atomic clocks keep time by monitoring the periodic vibrations of atoms in a rapidly moving stream. As the atoms move down a vacuum tube several feet in length, they emit and absorb radiation with a fixed, characteristic frequency that serves as the clock's pulse. However, the meandering of the individual particles within the stream spoils the regularity of their signal. This is precisely the kind of drawback Dehmelt wanted to avoid when he formulated his goal of capturing "a single atomic particle forever floating at rest in free space." A clock based on a single atom in a trap promises an improvement of accuracy over existing devices by a factor of at least a thousand and possibly much more.

Many groups of scientists in national laboratories, universities, and industrial research centers throughout the world are engaged in the quest to build better clocks, but they know that the road to success is long and arduous. Eventually the world's master clock will probably be an atom with an internationally approved official nickname kept in a gilded cage, but it will be years before we reach that stage.

The future of technological applications of Hans Dehmelt's single atom at rest in free space is assured, but more importantly it represents the realization of the defining dream of the atomic theory. As long as the atom was thought of as a miniature grain of sand, an object with a sharp boundary where it meets the emptiness of its surrounding space, the dream was vivid. But in the nineteen twenties quantum mechanics brought about a dissolution of the atom that caused this dividing line to become increasingly blurred; the electron shell has lost its palpability and taken on a spectral character halfway between being and noth-

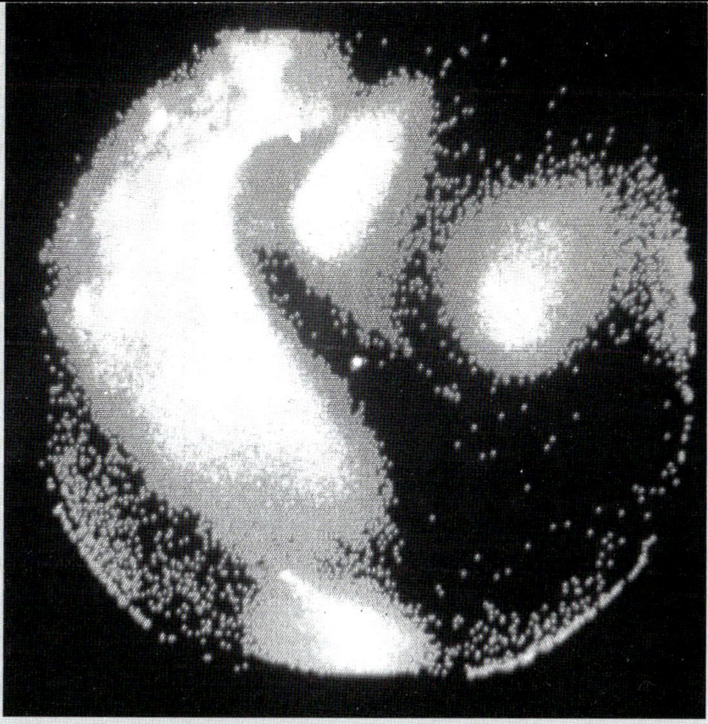

A charged mercury atom shows up as a little white dot in the center of the video monitor. The other shapes are merely reflections off the components of the trap; in the largest white crescent, for example, one recognizes part of the confining ring. COURTESY OF DAVID WINELAND

A color photograph of the barium atom Astrid, tiny and blue in the vacuum of a trap. COURTESY OF WARREN NAGOURNEY

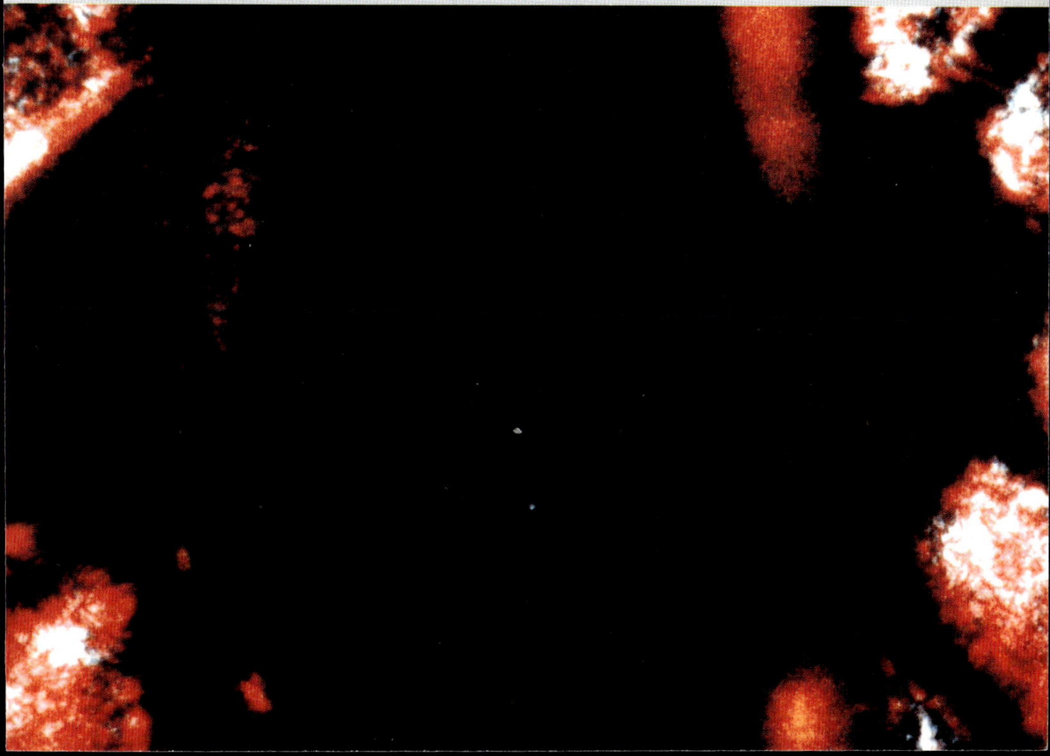

A scanning tunneling micrograph of benzene molecules lined up on a supporting metal surface. Each white lump is a single ring-shaped molecule, with a dark smudge indicating the hole through its center. COURTESY OF IBM RESEARCH

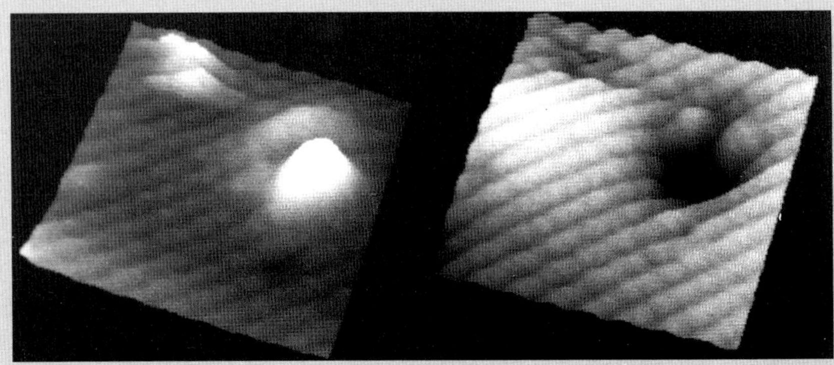

An oxygen atom on a gallium arsenide surface looks like a hill or a valley, depending on the direction of the electron flow from the surface to the STM probe. COURTESY OF RANDALL M. FEENSTRA

Oxygen atoms absorbed on a rhodium surface. COURTESY OF IBM RESEARCH

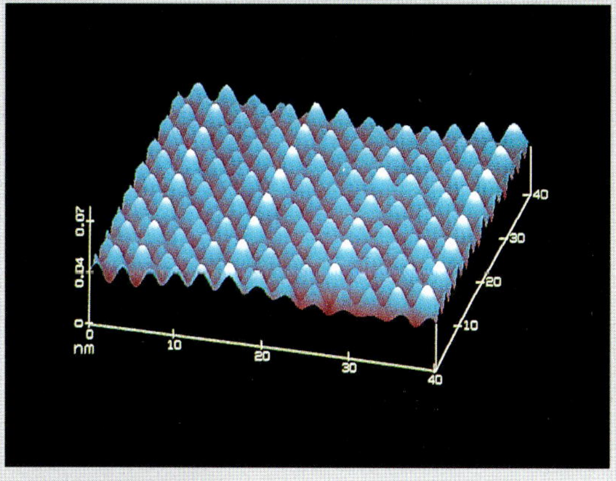

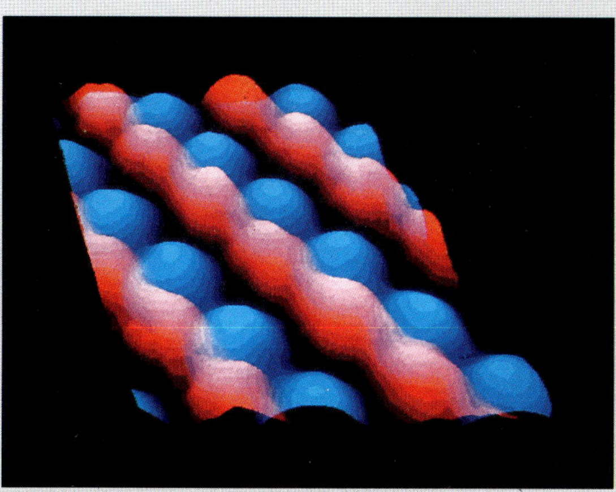

Rows of red arsenic atoms alternate with blue gallium atoms in a close-up STM picture of gallium arsenide. COURTESY OF IBM RESEARCH

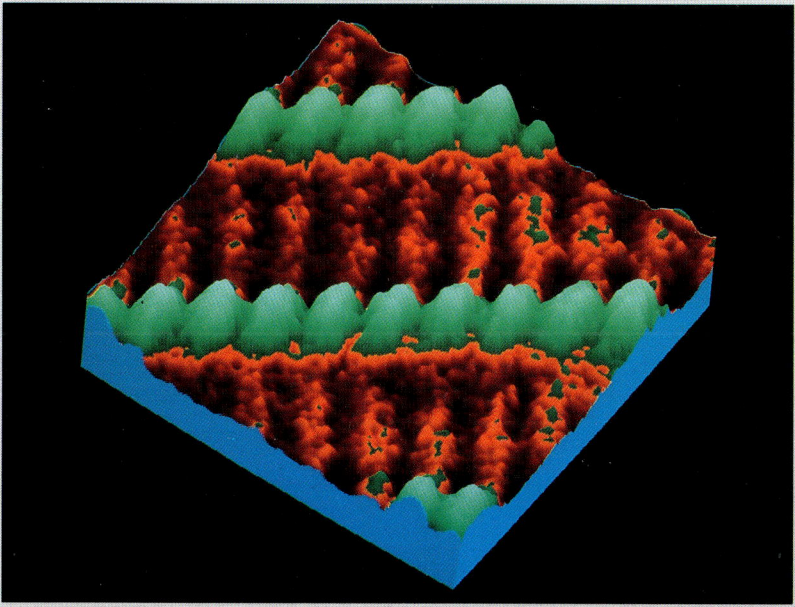

The rough surface of a complex liquid crystal on a graphite substrate. COURTESY OF IBM RESEARCH

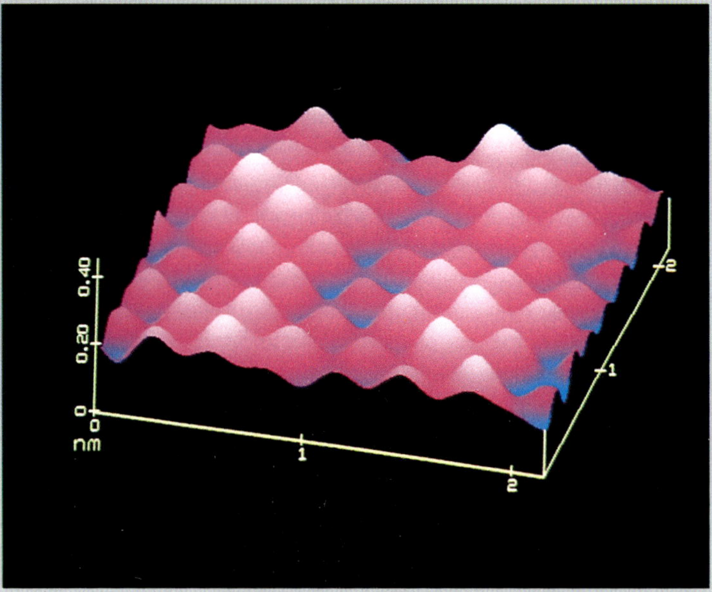

Ripples on a surface of a complex chemical–Tantalum Disulfide, an important lubricating material—bunch the atoms into hilly islands. COURTESY OF DIGITAL INSTRUMENTS

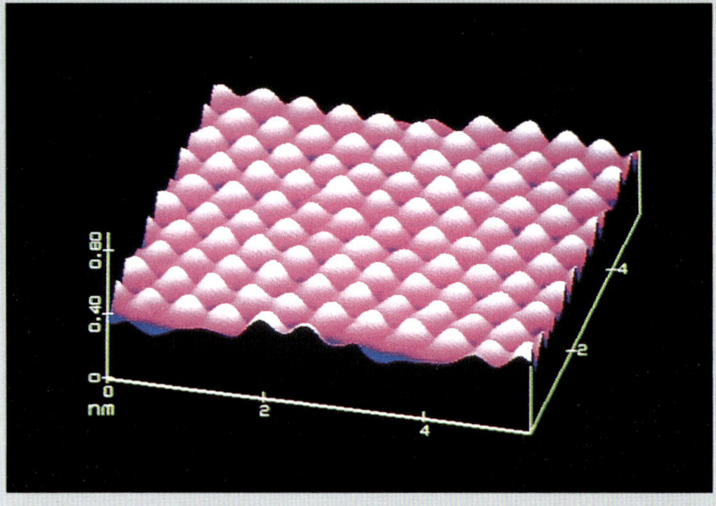

The surface of a single grain of table salt. COURTESY OF DIGITAL INSTRUMENTS

Six neon atoms at the edge of a one-atom step in the underlying platinum surface. The vertical scale has been exaggerated to make them appear taller. COURTESY OF *DISCOVER* MAGAZINE

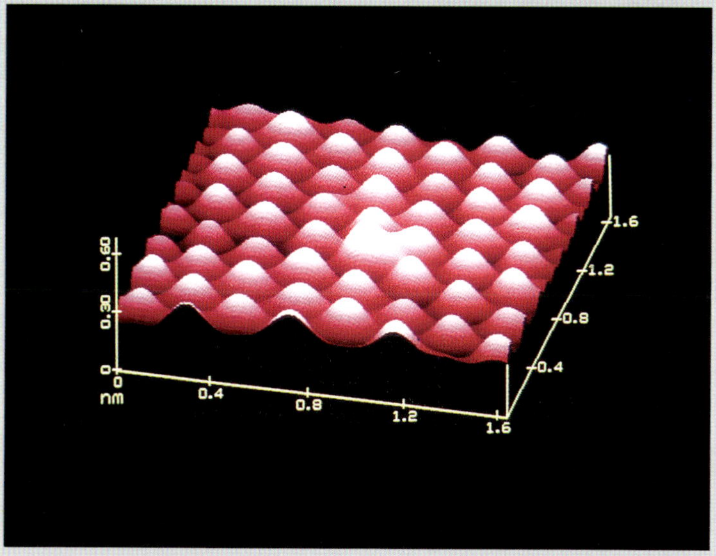

An atomic irregularity mars the surface of graphite. COURTESY OF DIGITAL INSTRUMENTS

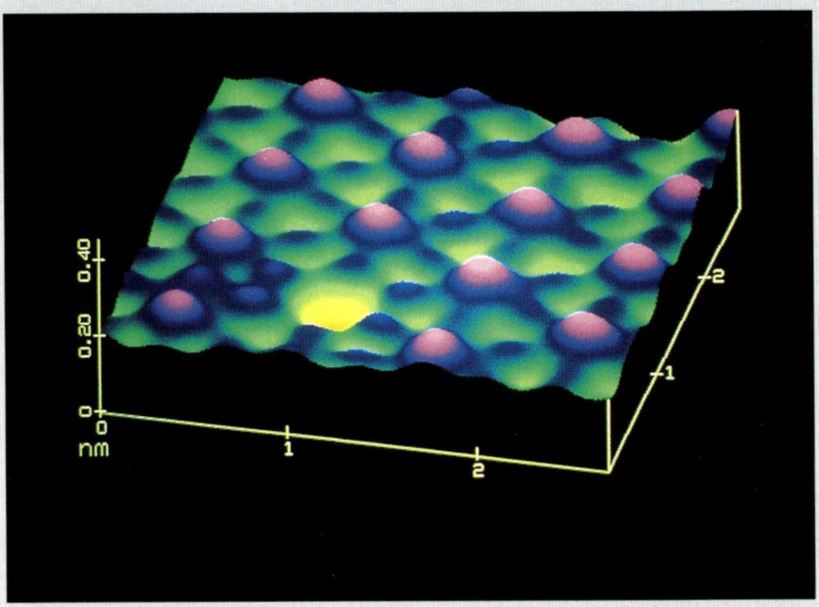

Iodine atoms are bonded to each other, but one is missing, leaving a gaping hole in the surface. COURTESY OF FRAN HEYL

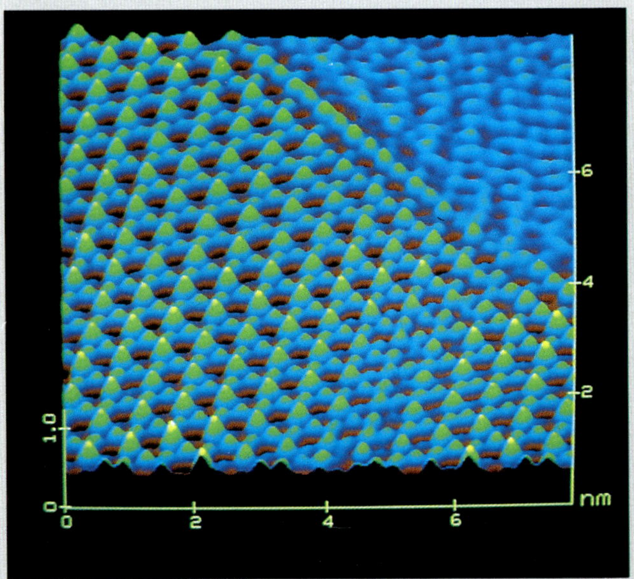

Iodine atoms on a platinum plate immersed in a liquid that contains copper atoms. The chemical interaction between the iodine and the copper powerfully effects the structure of the surface, accounting for the striking differences between this micrograph and the one pictured above. COURTESY OF DIGITAL INSTRUMENTS

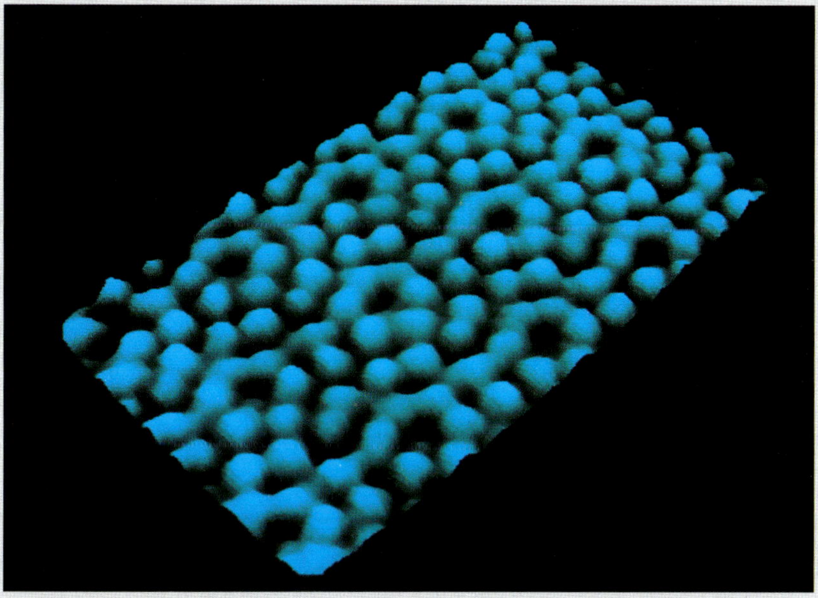

Silicon atoms form a hexagonal pattern on a surface. COURTESY OF IBM RESEARCH

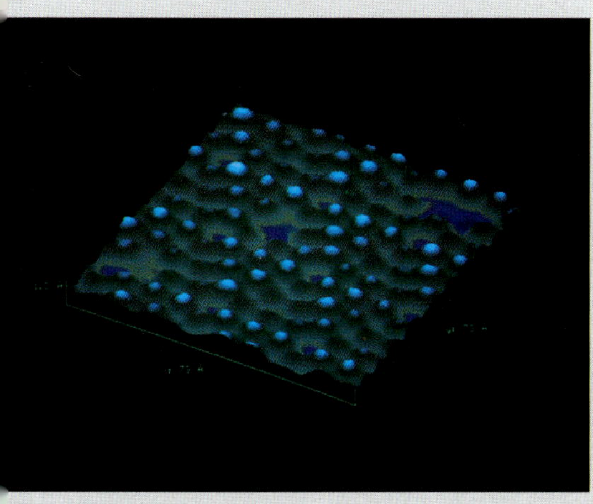

A more irregular silicon surface. COURTESY OF BURLEIGH INSTRUMENTS

The surface of silicon, magnified a billion-fold. COURTESY OF AT&T ARCHIVES

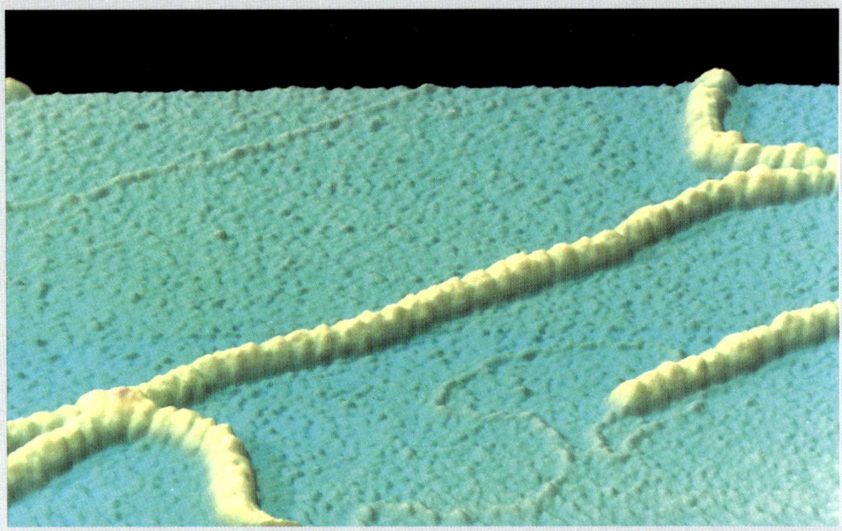

A DNA molecule, coated with a conducting film, snakes its way across a surface. COURTESY OF IBM RESEARCH AND *SCIENCE* MAGAZINE

A pyramid of germanium atoms. The Greek philosopher Democritus (c. 460–370 B.C.), in contemplating geometrical pyramids, might have imagined just such a structure. COURTESY OF MAX G. LAGALLY

ingness. At the same time a growing understanding of quantum mechanics has transformed the concept of the vacuum, the definite void beyond the atom's outer edges, into something more enigmatic than mere emptiness. Thus a true portrait of the atom in isolation must include a faithful depiction of the background, the quantum mechanical conception of free space, the vacuum.

7

Atoms and the Void

The universe consists mostly of vacuum; what few little bits of matter float around in its vast emptiness are hardly worth mentioning. In the immense regions of outer space between galaxies, astronomers have not been able to find any material at all. They acknowledge that there might be some that has escaped detection, but suspect that if you searched a volume as large as the Superdome, you would not come up with more than a single atom.

In our more immediate surroundings filled with solids, liquids, and gases, conditions are not much more crowded. A close-up of an atom would reveal that the nucleus, which carries 99.9 percent of an atom's weight, hovers in the center like a BB suspended in the Superdome; the rest is empty space, except for a few electrons that waft about the stadium like ghostly clouds of tenuous vapor. We and our world are made of pretty insubstantial stuff. It is surprising, therefore, how much thought and energy, not to mention money, scientists spend on trying to unravel the riddle of matter. Shouldn't they be worrying instead about the nature of the vacuum, which is by a wide margin the major constituent of the universe? Shouldn't they be thinking about nothing?

Some have been doing just that, in fact, and what they find is startling. The vacuum is a far busier place than it appears to be. Modern physics is on the verge of revealing it to be not merely a passive stage but an active participant in the processes of the material world. The vacuum, paradoxical as it sounds, interacts with atoms and may someday even become a working component of high-tech devices such as lasers. It is empty of matter, but filled with surprises.

Unlike the existence of matter, which cannot be called into question, the existence of the vacuum has been the subject of controversy since classical antiquity. The vacuum was initially introduced as an essential part of the atomic theory. "By convention there is sweet, by convention there is bitter, by convention hot and cold, by convention color; but in reality there are only atoms and the void," declared Democritus.

The vacuum of Democritus was a hypothetical concept required to make sense of the world as we perceive it. If matter were really the unbroken continuum we seem to perceive, how would a fish find room to swim forward, for example? Or, when a drop of milk dissolves in water, why does it seem to vanish into nothingness? Both puzzles are solved convincingly if there is a vacuum between atoms—to accommodate the front tip of the fish in the first case, and to hide the particles of milk in the second. But the void was rejected along with the atom by Aristotle for a variety of reasons. Among his arguments against the existence of the void, one is particularly ironic in hindsight. If there were a void, Aristotle claimed, all bodies would fall through it at the same rate, because they would encounter no resistance. Since this conclusion contradicted Aristotle's mistaken idea that heavy bodies always fall faster than light ones, he dismissed the void. If Aristotle had been able to watch, as millions of TV viewers did in 1971, how Apollo astronaut David Scott dropped a feather and a hammer from waist height on the airless moon and how they hit the ground at the same time, he might have learned to accept the void, and with it, perhaps, the atom.

Aristotle filled the vacuum with ether. This ether, not to be confused with the smelly chemical compound of the same name, was supposed to be a fine, universal substance that permeated all space, and even all corporeal bodies, but was impossible to detect. As an idea it had remarkable staying power, and survived Aristotle's original reason for invoking it. By the seventeenth century philosophical objections to the vacuum had been deflated; physicists had actually made a vacuum, or at least something close to it, with the help of the newly invented vacuum pump, which later became a crucial instrument in J. J. Thomson's discovery of the electron. Although the artificially produced vacuum was not perfect (even the best modern vacuum pumps haven't achieved that yet), it had become at least possible to imagine a space that was totally free of atoms. But the ether lived on.

Indeed, it flourished again to play an essential role in nineteenth-century physics. The proof by Thomas Young in 1803 that light consists of waves sparked a search for an invisible medium that could carry such waves. We are used to sound waves in air, water waves on the ocean, and even amber waves of grain, but we cannot imagine waves of void. Yet light, unlike sound, can travel through an apparently empty space—it reaches us from the sun and the stars, for instance. Space, physicists reasoned, could not really be empty. It must be filled with ether.

This ether, however, had more definite physical properties than Aristotle's. It was known that sound waves move faster in a denser medium, such as water, than in a softer one, such as air. Since the speed of light is so tremendously high—186,000 miles per second—the ether had to be exceedingly firm, even solid. And yet planets move through it without encountering any detectable resistance. It was strange stuff indeed, this ether, at the same time denser than steel and thinner than air, but the physicists of a hundred years ago could see no way to dispense with it.

From 1887 on, however, they began to have doubts about its reality. In that year the American physicists Albert Michelson and Edward Morley conducted an experiment to prove the existence of the ether. If the earth were moving through a stationary ether, they reasoned, it should feel an ether wind. When light was bucking this headwind, it should travel more slowly than when it was carried along downwind. Michelson and Morley tried to measure this difference, but found none: The speed of light was the same in either direction. The result was completely unexpected and shook the foundations of the ether theory. Physics, it seemed, made no sense without the ether, nor with it.

In 1905 Albert Einstein cut through the dilemma in his characteristically blunt way. In the introduction to his first paper on the special theory of relativity he simply declared the ether hypothesis to be "superfluous." At the age of twenty-six he boldly disposed of a 2,400-year-old physical concept with a stroke of his pen. To critics who might object that waves need a medium to carry them, he replied, in effect, "That may be true for some waves, but for light it just doesn't happen to be so." The impressive soundness of the rest of his theory, together with the negative result of the Michelson-Morley experiment, finally put an end to the ether.

The vacuum, thus cleansed, remained empty for a quarter of a century, but then it began to fill up again. This time the culprits were neither matter nor ether but products of the quantum theory. After its initial successes in the theory of matter, quantum mechanics was applied to the more subtle problem of the vacuum. In the late nineteen thirties and early forties a number of scientists throughout the world, including Richard Feynman, perfected the conception of the modern, dynamic vacuum. Even though it is empty of matter, it is buzzing with energy and hidden activity. The modern vacuum represents a compromise between the opinions of Democritus and those of Aristotle: The former was right to insist that the world consists of atoms and the void, and the latter when he claimed that there is no such thing as true and absolute emptiness.

The two novel features introduced into the vacuum by quantum mechanics are vacuum fluctuations and vacuum polarization. Both terms serve notice that the dynamic vacuum is something new: True nothingness can neither fluctuate nor exhibit polarity. Both phenomena are based ultimately on the uncertainty principle, that basic tenet of quantum mechanics that forbids the simultaneous and certain determination of the position and speed of a particle or, in the words of Tom Stoppard, "If you know where it is, you don't know what it's doing, and if you know what it is doing, you don't know where it is."

One consequence of the uncertainty principle is the so-called zero-point energy of mechanical systems. If, for example, two atoms are joined together to form a molecule that resembles a stiff spring with a weight at each end, they will naturally vibrate back and forth along their common axis. In fact the vibration can never be completely eliminated. There always remains a last irreducible quivering, known as the zero-point motion, a tremor like that of an aspen leaf in the wind. It ensures that the uncertainty principle is satisfied even when the molecule is practically at rest, and has been observed experimentally. The zero-point energy caused by this motion blurs the sharpness of the energy levels of the molecule, which in turn results in a reduction in the purity of the colors of light emitted by the molecule. But the vibration of atoms and molecules represents only one type of oscillation, and is irrelevant to the vacuum; there is another type involving electromagnetic fields that is not.

According to James Clerk Maxwell's nineteenth-century theory of electricity and magnetism, light consists of oscillating electrical and magnetic fields. The oscillations of these fields are just as much afflicted with zero-point energy as atoms are and therefore can never quite vanish. According to the quantum theory, then, the electromagnetic field is never entirely absent, even in a completely dark vacuum. Random electromagnetic fields always fluctuate gently in such a space, and every fluctuation carries its own zero-point energy.

In a small volume of, say, a cubic inch, the energy of each electromagnetic fluctuation is very small, but because there is an infinite number of possible waves, the sum total of all their zero-point energies turns out to be infinite. Paradoxically the impossibly dense ether has been replaced by an infinite energy density pervading the entire universe. This mind-boggling conclusion inspires inventors to propose all kinds of schemes for solving the world's problems by harvesting energy from the vacuum, but most physicists dismiss them as fanciful. Only energy differences matter in physics, not absolute energies. Only when one system exchanges energy with another one, either by giving some away or by receiving some, does a physical process unfold. The millions of dollars in your bank's vaults are of less significance to you than the sixteen-dollar check you have to cover next week or the five-dollar bill you extract from the money machine. Just as you ignore your bank's total assets when you conduct your everyday finances, physicists simply ignore the infinite energy of the vacuum required by the rules of quantum mechanics.

Besides the electromagnetic fluctuations of the vacuum, quantum mechanics predicts an even more exotic phenomenon, called vacuum polarization. Occasionally an electromagnetic fluctuation carries enough energy to materialize spontaneously into a pair of particles: Energy turns into mass without human help. Most often the particles turn out to be an electron and its oppositely charged antiparticle, a positron. Together they carry no net electrical charge, and their masses, when converted into energy according to Einstein's famous formula, add up to the energy of the fluctuation that gave birth to them. All of a sudden, for a brief moment, a negative electron, together with its positively charged sibling, pops up out of nowhere, and in an instant they annihilate each other and vanish without a trace. If there happens to be a strong positive electric charge nearby, the electron will be attracted to it, and the positron repelled, so that during its brief lifetime the pair line up like a compass needle. In this way the vacuum becomes momentarily polarized.

The dynamic vacuum is like a quiet lake on a summer night, its surface rippled in gentle fluctuations, while all around, electron-positron pairs twinkle on and off like fireflies. It is a busier and friendlier place than the forbidding emptiness of Democritus or the glacial ether of Aristotle. Its restless activity is utterly fascinating to physicists and invites speculation about its nature and even its potential usefulness. As a theoretical conception the dynamic vacuum holds great appeal, but whether or not it had any physical validity could only be decided in the laboratory.

The drama of the experimental investigation of the modern vacuum took place in three acts culminating in a kind of marriage of atoms with the void. First, when theoreticians like Richard Feynman returned from the secret bomb laboratories of World War II to resume their careers in basic research, they were delighted to learn of experiments that showed the microscopic interstitial vacuum inside the atom to be as dynamic as they had come to believe. Just a few months later a Dutch industrial physicist discovered quantum mechanical effects in the ordinary macroscopic void in an empty vessel. Finally, within the last five years, the most recent phase of experimentation has brought to light the nature of the interaction of the dynamic vacuum with individual atoms. Nothingness has merged with matter in one indivisible entity.

The first act took place in the old Pupin Laboratory of Columbia University, where I later learned physics as an undergraduate. In 1947 Willis Lamb, a quiet, scholarly Californian who had joined the faculty just before the war at age twenty-five, set out to measure the energy levels of the hydrogen atom with unprecedented accuracy. By that time its detailed structure had been calculated from Schrödinger's theory, with corrections due to Einstein's relativity, and theoreticians thought they understood it completely.

Lamb realized that certain transitions between levels could be induced by microwave radiation rather than by light. Microwaves are common in the kitchen today, but during World War II they were new, and their chief application was radar. Lamb had spent the war years perfecting techniques for generating and detecting microwaves and was happy to be able to turn this practical expertise to use in the pursuit of pure science. When he measured the step sizes of hydrogen's energy staircase by recording the specific frequencies of microwaves that were absorbed by hydrogen gas, he found that they differed from the calculated values by one part in a million. Confident of the high quality of his work, he urged the theorists to explain the discrepancy, which later became known as the Lamb shift.

His theoretical colleagues quickly rose to the challenge by attributing the phenomenon to the effect of the vacuum. They realized that as the lone electron of the hydrogen atom circles its nucleus, it is exposed to fluctuations of the vacuum that cause it to jiggle faintly and change its energy by a minute amount. To make matters more complex, the polarization of the vacuum also contributes slightly to the Lamb shift, but to a much lesser degree, because it only results from those rare fluctuations that happen to carry enough energy to produce electron-positron pairs. The precise agreement between the experimental value of the Lamb shift and its theoretical explanation showed that Lamb had indeed measured a change in an electron's energy caused by fluctuations and polarization of the atomic vacuum in which the constituents of the atom reside.

The setting of the second act of the drama of the dynamic vacuum was an industrial laboratory. In 1948 Hendrik Casimir, then director of the research laboratory of the Phillips Electric Company in Eindhoven, Holland, was working on the question of how electrically neutral particles of matter suspended in a liquid affect each other. Casimir is a physicist's physicist: his name is as respected within the profession as it is unknown outside it. Although he spent the majority of his career wrestling with the nitty-gritty problems that crop up on the factory floor, his academic colleagues throughout the world value his knowledge of quantum mechanics at its profoundest levels. Casimir's explanation of what brought him to the problem of interatomic forces and then, by serendipity, to the vacuum, demonstrates how applied research, properly performed, can serve pure science: "Suspensions are frequently used in the electronic industries," he wrote forty years later, "when surfaces have to be covered with fine powders, as in cathode-ray tubes and in luminescent lamps, and it was an accepted policy of the research laboratories to try to really understand empirical procedures and not be satisfied with a recipe that worked well in practice but was not understood."

Using quantum theory and the known laws of electricity and magnetism, Casimir carefully calculated the mathematical expression for the force of attraction between two neutral atoms—a very weak force that is the net result of the mutual repulsions and attractions of their electrically charged components. The calculation was long and tedious, but the final result turned out to be a very simple formula. It was as if he had covered twenty pages with complex computations and ended up with something like 2 + 2 = 4. Surely, he thought, there must be a quicker way to reach the same conclusion, but he couldn't think of one. On a visit to Copenhagen he happened to mention this suspicion to Niels Bohr, at the time the reigning patriarch of quantum mechanics. As Casimir recalls, Bohr muttered something like "The force must be a manifestation of zero-point energy" and promptly dropped the subject. It was an intuitive oracular pronouncement of the kind Bohr was famous for, and Casimir pursued it to its startling conclusion.

In order to simplify the calculation, he considered the force between two large, parallel, uncharged metallic plates in a vacuum, in place of two spheres or two atoms. He assumed that no gases would be present to push on them, that their mutual gravitational attraction would be negligible, and that they'd be cool enough for thermal radiation to be insignificant. Then, if the plates were both electrically neutral, nothing in classical physics would produce a force between them. Following Bohr's clue, however, Casimir found that the plates modify the dynamic vacuum between them and that this change, in turn, causes a force of attraction where none was expected.

To understand how this could occur, consider a flat, vertical wall of steel, several miles long, in the middle of the ocean. Imagine that it is attached to the ocean floor and protrudes above the water's surface by several feet. Waves of all kinds, from the tiniest ripples to the longest rollers, pummel it from the left and exert a push on it. At the same time similar waves push from the right, so the forces balance and cancel each other out. The wind, which causes the waves in the first place, is not under consideration here; in fact it would be best to think of the air being completely still. Now consider another flat wall, parallel to the first and ten feet away from it. In the space between the walls the water ripples just as merrily as on the outside, but long rollers perpendicular to the walls cannot grow there. The reason is simple: Since the total amount of water between the walls cannot change, the waves in the gap must have both crests and troughs. If the crest of a wave happens to be longer than the ten-foot separation between the walls, there clearly isn't enough room for a trough as well, so waves with long wavelengths simply do not exist between the walls. The net result is that there are long waves on the outside that are not counterbalanced on the inside and hence push the two parallel walls together.

Metallic plates exposed to vacuum fluctuations behave very similarly: The fluctuations in the vacuum, outside and inside the space between them, combine to force the plates toward each other. (The effect shows up only for metal plates because they are impenetrable to electromagnetic radiation. Glass plates, for example, are transparent to light and radio waves and therefore powerless to exclude vacuum fluctuations.) A simple calculation led Casimir to the astonishing prediction that the dynamic vacuum causes two neutral metal plates to attract each other. His claim was soon put to the test. A researcher at the Phillips Laboratory rigged up a delicate spring mechanism to show that plates measuring a few centimeters square and separated by a micron—a millionth of a meter—attract each other with a force equal to the weight of a mosquito, just as predicted by Casimir's formula. The force may be puny, but the fact that it is caused neither by gravity nor by electricity but by the vacuum itself has continued to fascinate physicists for over forty years. In 1987 Casimir proudly cited two recent bibliographies totaling 483 references to the effect that bears his name.

Neither the Lamb shift nor the Casimir effect suggest any practical uses for the vacuum. The former is a natural property of atoms, a consequence of the microscopic vacuum that cannot as yet be switched on or off. The latter, being macroscopic, is indeed open to manipulation, but its magnitude is too small to be significant. Although many details and refinements of the Lamb shift and the Casimir effect have been worked out in the last four decades, the dynamic vacuum has not yet risen to the level of common consciousness the way the atom has.

During the last decade the curtain has been raised on the third act. The story of the vacuum is about to enter a new phase. The latest developments concern the behavior of individual atoms in a vacuum between plates and thus combine features of the microscopic vacuum responsible for the Lamb shift with those of the macroscopic Casimir effect. The vacuum influences the emission and absorption of radiation by atoms. An electron that has absorbed some energy by exposure to heat or light or an electrical spark soon drops back to its original lower position on the energy staircase. The attendant radiation of light or radio waves is called spontaneous emission, and, until recently, it was thought to be an intrinsic property of the atom, as inevitable as the explosion of a bomb with an automatic timer fuse.

But where does the radiation emitted by an atom go? Presumably it escapes into the surrounding vacuum, unless the vacuum isn't prepared to receive it. The Casimir effect suggests a way to ensure such a condition: If the vacuum happens to be enclosed in a narrow metallic cage, and if the radiation to be emitted has a wave length too long to fit into that cage, the vacuum will not be receptive, and the spontaneous emission of radiation will not occur.

This is the crux of the modern exploitation of the dynamic vacuum. Spontaneous emission is not an intrinsic property of an isolated atom but the result of an interaction between the atom and the vacuum. Since the quantum mechanical nature of the vacuum can be changed by adjusting the geometry of its container (for example, by varying the size of the gap between two plates), spontaneous emission can he inhibited artificially. Conversely, spontaneous emission is enhanced when the atom is given a container made especially receptive to a certain wavelength. This phenomenon is called resonance and is familiar from music: Different lengths of organ pipes sustain, or resonate to, different pitches, which correspond to specific wavelengths of sound. Thus the process of spontaneous emission, which was once thought to be an invariable attribute like mass or charge, can be controlled by the modification of the vacuum.

One of the simplest and most convincing experiments to demonstrate a modification of this kind was performed in 1985 by a group working under Daniel Kleppner, an urbane, silver-haired professor who seems just as much at home at the Collège de France in Paris as he is at MIT in Boston. Kleppner is a renowned teacher and graceful writer whose trenchant observations of the world of physics frequently enliven the pages of the journal of the American Physical Society. His pioneering contributions to the art of manipulating atoms and the vacuum were appropriately honored in 1986 with the prestigious Davisson-Germer Prize, which commemorates the first empirical demonstration of the wave nature of atomic particles. His writing as well as the design of his experiments display a quality that distinguishes them from the run-of-the-mill: Daniel Kleppner has style.

To modify the vacuum, he and his team built a device consisting of two parallel copper plates, analogous to the imaginary walls in the ocean, that were approximately twenty centimeters in length and separated by a millimeter gap whose width could be varied very precisely. A beam of cesium atoms was excited to a state of high energy and directed between the plates along the entire length of the gap. Just before they entered this channel, they passed a magnet that oriented them in such a way that their spontaneous radiation, if it were to occur, would be produced in a direction perpendicular to the plates—like ocean waves that travel perpendicular to the walls—because radiation emitted parallel to the metal plates had no cause to be inhibited at all. At the far end of the plates a radio receiver tuned to the frequency of emission monitored the number of atoms that made it through without having undergone spontaneous emission.

When the gap was wide, most of the atoms lost energy by spontaneous emission during their long trip through the channel. The monitor showed a steady, weak signal, which revealed that few of the excited cesium atoms remained intact. But when the separation was made narrower and approached a value equal to half a wavelength, so that the gap could accommodate no more than a single crest or trough, the monitor signal shot up to four times its original value. This was a clear indication that many more of the excited atoms were getting through, because spontaneous emission had been inhibited, and that the vacuum around them had been modified.

Other scientists are now working toward the complete suppression of spontaneous emission, an achievement that will lead to dramatic advances in laser performance. Since lasers depend on controlled emission, which is spoiled by undisciplined acts of spontaneous emission, suppression of the latter improves the laser action. Other devices that are made possible by artificial control over the emission of radiation haven't even been dreamed of yet.

From a more fundamental point of view, experiments like Kleppner's demonstrate that the atom physically interacts with its surrounding vacuum and through it with the walls of its container. A theory that deals with the atom without referring to the vacuum, and a description of the vacuum that fails to consider its boundaries, are incomplete. In fact, modern versions of quantum mechanics put electrons, electromagnetic fields, and their mutual interactions on an equal footing and thereby provide a complete description of the effects of the vacuum on atoms.

However, all theories still persist in drawing a sharp distinction between particles and fields, and this dichotomy lies at the root of the paradoxical nature of the vacuum. The vacuum's strange properties—infinite energy density, fluctuations, polarization, the Lamb shift, the Casimir effect, the inhibition of spontaneous emission—stem ultimately from the artificial separation of charged particles from the electrical and magnetic fields that surround them. Albert Einstein understood this deficiency of present-day quantum mechanics long ago. At a seminar at the Princeton Institute for Advanced Study in the 1940s, he said, "I feel it is a delusion to think of the electron and the field as two physically different, independent entities. Since neither can exist without the other, there is only *one* reality to be described, which happens to have two different aspects; and the theory ought to recognize this from the start." Such a theory does not exist yet, but if it is discovered, it will clear up the vacuum once again.

If electrons and their fields are regarded as parts of the same underlying reality, the theoretical model of the atom will change profoundly. At present the boundary of the atom is blurred by the indistinctness of the electron's wave function. Nevertheless, since the probability of finding the electron drops off precipitously at large distances from the nucleus, the wave function appears to have a boundary resembling the edge of a wad of cotton. The electric field of the electron, on the other hand, reaches out far beyond the atom into the vacuum and from there to other atoms and even distant walls. If this influence becomes part of the atom, our mental image of the atom, which began as a firm kernel and evolved into a ball of cotton, will puff up into an all-pervading cloud.

Even without the realization of Einstein's radical suggestion, the exploration of the dynamic vacuum has produced one of the principal insights of modern physics—the proposition that contrary to its fragmented appearance, the world at its most basic level is a connected unit. The walls of a vessel, the vacuum it contains, and the atom in its center can no longer be imagined as separate entities. Like the distinction between subject and object in philosophy, and the separation of mind from body in medicine, the division of the world into matter and vacuum turns out to be illusory. By intuition and reason Democritus penetrated to the very heart of nature when he discovered that reality is to be found in atoms and the void, but he could not have foreseen what we are learning today: that stars and atoms and the vacuum are all part of a single, seamless whole.

The merging of the diffuse electron wave with the random rippling of the surrounding vacuum is rendered even more indistinct by the superhuman speed at which atomic events occur. The time it takes an atom to negotiate the length of Professor Kleppner's elegant apparatus at MIT amounts to mere billionths of a second, and while the intervals between quantum jumps of an atom in a trap are long enough to be seen with the naked eye, the duration of a jump itself is too short to be measured. The atom is not a static structure but a dynamic mechanism in constant interaction with its equally dynamic environment. It is not a grain of sand but a wave-tossed buoy blinking from afar. If we want to understand it, we must look beyond still pictures and record the action in a movie.

8

Atoms in Action

The headline in the *Honolulu Star-Bulletin and Advertiser* for the thirteenth of May 1990 read: CATS—SPINS, PHASE SHIFTS, AND LANDING ON YOUR FEET. Underneath it a blowup of an Associated Press wirephoto showed a sleek striped cat in the act of coming in for a landing after a considerable fall, its feet and tail splayed out in all directions. A little to the left a young man in jeans and a striped shirt kneels on the ground in front of a barn, evidently trying to follow the contortions of the twisting feline body. His face mirrors the cat's expression of intense concentration. A caption below the photo identified the cat as Sam, who, it said, always lands on his feet.

The ability of falling cats to right themselves in midair and land on their feet has been a source of wonder for ages. Mischievous children throw the family tabby off the garage roof to watch this thrilling trick, and biologists regard it as an example of adaptation by natural selection, but for the physicists it borders on the miraculous. Newton's laws of motion guarantee that the total amount of spin of a body cannot change unless an external torque speeds it up or slows it down. If a cat has no spin when it is released, and experiences no external torque, it ought not to be able to twist around as it falls. (Air resistance is much too feeble a force to provide the necessary torque.)

In the late nineteenth century this puzzle excited the curiosity of professional and amateur physicists. In time the dilemma was resolved, but the apparent conflict between the evidence of our eyes and the conclusions of our brains is so startling that the mystery of the falling cat never really goes away; like a cat it has many lives. Every dozen years or so a journal devoted to the teaching of physics, or a book about the wonders of nature, or a newspaper like the *Honolulu Star-Bulletin and Advertiser,* brings it up all over again and sets off a fresh round of discussion. Every generation has to work its own way through the conundrum.

In the speed of its execution the righting of a tumbling cat resembles a magician's trick. The gyrations of the cat in midair are too fast for the human eye to follow, so the process is obscured. Either the eye must be speeded up or the cat's fall slowed down for the phenomenon to be understood. A century ago the former was accomplished by means of high-speed photography, with equipment now available in any drugstore, but in the nineteenth century the capture on film of a falling cat constituted a scientific experiment. What was needed was a shutter fast enough to stop the action, a camera capable of taking several pictures in the short time that the cat is in the air, and film of sufficient sensitivity to work under such demanding conditions.

The experiment was successfully carried out by one Monsieur Marey who described his results in a paper presented to the Paris Academy in 1894 and published in the journal *La Nature.* Two sequences of twenty photos each, one from the side and one from behind, show a white cat in the act of righting itself. Grainy and quaint though they are in comparison with the modern wirephoto, they leave no doubt that the cat was dropped upside down, with no initial spin, and still landed on its feet. (Its face is indistinct, but its body language conveys a feeling of offended dignity at the end of the experiment.) Careful analysis of the photos reveals the secret: As the cat rotates the front of its body clockwise, the rear and tail twist counterclockwise, so that the total spin remains zero throughout the trip, in perfect accord with Newton's laws. Halfway down, the cat pulls in its legs before reversing its twist, and then extends them again, with the desired end result. The lesson, for the physicist, is that while no body can acquire spin without a torque, a flexible one can readily change its orientation, or phase. Cats know this instinctively, as do champion divers and dancers, but scientists could not understand it until they managed to increase the speed of their perceptions a thousandfold.

The mystery of atomic motion presents a similar challenge. We readily perceive both the initial and the final configurations, but the events in between happen so quickly that we cannot follow them.

Consider a simple chemical reaction, say that of hydrogen combining with carbon dioxide to form hydroxide and deadly carbon monoxide, $H + CO_2 \rightarrow OH + CO$ in symbols. How does it work? Does the carbon dioxide molecule shed an oxygen atom for the passing hydrogen atom to catch? That seems improbable, because if carbon dioxide spontaneously decayed into carbon monoxide, we would die of our own exhalations. If, on the other hand, hydrogen hits carbon dioxide, the system becomes, at least temporarily, a congeries of four atoms—hydrogen, carbon, and the two oxygens. Then how do they reach their final outcome? In what manner do they twist and turn as they reassemble? Do they briefly form some new molecule hitherto unknown to chemists? If so, what is its shape? How long does it live? These are questions as urgent to modern chemists as the motion of a falling cat was to physicists during the Victorian age.

Ahmed Zewail, a professor of chemical physics at Caltech, and his colleagues have been studying rapid chemical reactions since 1980. When the Egyptian-born scientist first began this research, he was faced with a design problem reminiscent of Monsieur Marey's. A cat dropped from a height of several feet acquires such a high speed that in a thousandth of a second it travels a quarter of an inch. That is about the maximum distance the cat may be allowed to move during one exposure without blurring the image beyond recognition. The shutter speed must therefore be a millisecond or faster. The speed of falling objects, and the tolerable motion for proper exposure set the scale for Marey's technical requirements. What set the scale for Zewail's?

How fast do atoms interact? If the question were cast in more human terms, with the word *atoms* replaced by *lovers,* we would separate it into two parts—the times allotted to courtship and to consummation. The time scale of the former varies enormously and is dependent on many personal and societal factors, whereas the duration of the latter is much briefer and more uniform. In the same way, a chemical reaction requires first that the atoms come into contact in the right places and under the right circumstances. This can take a lot of time, as when sugar dissolves in iced tea, or very little, as in an explosion. The second phase, the actual process of two atoms undergoing chemical transformation, is considerably more rapid and depends on the atoms themselves, not their surroundings. An understanding of the preliminary meandering can teach us something about the flow of heat, about pressure and density, and about the physical properties of fluids and solids; the final step, the intimate act of molecular mating, helps to reveal the intrinsic nature of the atoms themselves.

Until recently, however, there was no way of even coming close to measuring how long two atoms took to complete this final step. To estimate the scale of time involved, one can temporarily ignore what has been learned about atoms in the last eighty years and reach all the way back to 1913, to the picturesque, though incorrect, Bohr model of the atom. In this context the positively charged molecule of hydrogen gas consists of two nuclei bound together by a single shared electron. This molecule may be said to have formed when the two separate nuclei have approached each other and when the single electron has performed at least one complete circuit around them. The time required for an electron to make this journey furnishes at least a rough idea of the scale of time involved in a molecule's birth.

From the known speeds and orbits of electrons in Bohr's model, the approximate time it takes them to circle a nucleus is easily calculated. The answer is a span so unimaginably brief that ordinary English fails. New words have to be coined to describe it, and so it turns out that atomic interaction times are measured in femtoseconds.

A femtosecond is 1/1,000,000,000,000,000 or one quadrillionth of a second. In its struggle to keep up with the rush of discovery, scientific nomenclature progresses by factors of a thousand, not bothering to pause at intermediate steps of ten or a hundred. When Monsieur Marey designed his camera to shoot pictures in a millisecond, the fraction he had to deal with was one part in a thousand, a ratio that is still within the limits of human comprehension. The next division is the microsecond, a millionth of a second, a term that trips off the tongue with the greatest of ease but that can scarcely be grasped. A million dollars is no longer an exceptional fortune, but to imagine a millionth is more difficult. A second is approximately a millionth of a fortnight, and since both of those units of time are intuitively meaningful, a microsecond has some hope of being understood by analogy. Next is the nanosecond, a billionth of a second. Today billions are bandied about without apology in discussions of the costs of weapons systems, of the number of stars in the sky, and of people in the world. The nanosecond, at the other end of the scale of magnitudes, is named after the Greek word for "dwarf" and is to the second as a second is to thirty years. The prefix *nano-* is beginning to enter the world of engineering, to measure distance, not time. Cornell University, for example, created a new laboratory in 1987 and called it the National Nanofabrication Facility. Its mission is to conduct advanced research in the development of structures with dimensions measured in nanometers for use in biology, chemistry, and electronic engineering.

A thousand times shorter than the nanosecond is the picosecond, a trillionth of a second. The word *trillion* is sometimes found in news headlines about the annual U.S. budget, but it has little meaning for most people. A trillion seconds is longer than recorded history, too long to be comprehensible. The prefix *pico-* comes from the Spanish and Italian word for "small," whence the instrument called the piccolo. The final step is the femtosecond, a quadrillionth of a second, a second divided by a one followed by fifteen zeroes. Having exhausted the Latin *(milli),* Greek (*micro* and *nano*), and Romance languages *(pico),* English makes one of its rare excursions into Scandinavian to borrow *femto,* the root of the word for "fifteen." Neither a quadrillion nor a femto-anything has a chance of evoking mental images, so the statement that chemical transformations last femtoseconds is only a curious abstraction. (The spatial analog of the femtosecond fares a little better. The dimensions of atomic nuclei are measured in femtometers, but since the study of nuclear sizes began with Ernest Rutherford long before the word *femtometer* was coined in 1970, an older term has precedence. Nuclear physicists call a femtometer a fermi, in honor of the universally admired Italian-American physicist Enrico Fermi, who in 1942 achieved the first self-sustaining nuclear chain reaction and thereby ushered in the atomic age. By cleverly abbreviating both the femtometer and the fermi as *fm,* physicists can continue to comply with international nomenclature yet privately endow a clumsy technical term with warm human connotations.)

Ahmed Zewail and his colleagues set themselves the daunting task of designing an instrument to monitor reactions at femtosecond shutter speeds, 10^{12} times faster than the speed with which Monsieur Marey took pictures of cats. The new device had to be based on the laser, which, since about 1985, has become capable of producing pulses of light as short as a few femtoseconds. A hundred years ago chemical research was centered on the synthesis of organic dyes from coal tar. Today the tools of femtochemistry are pulsed, tunable lasers whose luminous elements are organic dyes. In the cycle of scientific research, the pure science of yesterday bears fruit in today's technology, which in turn leads to new fundamental discoveries.

To visit Zewail's laboratory at Caltech, in a room he inherited from Linus Pauling, is to enter a magic kingdom. Like David Wineland's in Boulder, it is crowded from wall to wall with optical tables. Scores of optical devices, each piece a separate precision apparatus about the size of a small camera, are mounted on the tables in rows. The place is dimly lit, but the laser beams themselves are visible as brightly colored threads that impose order on the confusing scene, like gleaming tracks through an elaborate, albeit motionless, model railroad display at Christmas. Among the lens holders, prism tables, and shutters, painted black to reduce reflection and equipped with knurled adjustment screws, the laser beams gleam red and green and orange, enter black boxes and unexpectedly emerge on the other side in a different color, fan out in diverging directions before reassembling again, reflect off mirrors and crisscross the room. In several places looping plastic tubes supply the multicolored organic dyes to the lasers. Everything looks clean, precise, pristine.

If, in all this icy perfection, there is a place where molecules are caught in the act of tumbling through space, it is not discernible. The furious sounds of screaming cats and cursing assistants, the funky smell of overheated lights and human sweat, the frantic confusion that must have filled Monsieur Marey's studio are nowhere to be found. And yet, as Professor Zewail points out, the basic principle of his experiments bears a close similarity to old-fashioned high-speed photography. Both techniques separate a continuous motion into a series of snapshots, which can later be reassembled into a reasonably accurate representation of continuous motion. While Marey's experiment aids in our understanding of Newtonian physics, Zewail's helps to illustrate quantum mechanics by recording the formation of molecules. The goals in both cases are similar; only the scales have changed, from milli- to femtoseconds and from centi- to nanometers.

The way to produce laser flashes, or pulses, in the femtosecond range resembles the trick by which piano tuners make their living. Consider two short bursts of sound, one from a tuning fork and the other from a piano string with a very slightly different pitch. If the two sources are struck simultaneously, the result will not be two pitches but a single note growing louder and softer in sinuous succession. The distinctive characteristic of these modulations, which are called beats, is that they grow longer and farther apart as the two constituent tones approach each other in pitch, disappearing altogether when the pitches coincide. Each beat, from silence to maximum loudness to silence again, is a pulse of one single frequency and lasts for a much shorter span of time than the original bursts of sound. Indeed, as the two original pitches diverge, the beats become even shorter, so that short pulses can be created from longer ones. In the case of lasers pitch is replaced by color. Picosecond pulses of slightly different colors are combined to give beats that last mere femtoseconds. Thus the fastest laser pulses known to science are conceptually as simple as a piano tuner's beats.

The short pulses fulfill two very distinct functions. In Marey's experiments the preparation of the initial state was performed by the assistant who held the cat upside down and who presumably called out when he was ready to drop it. In a chemical reaction like the conversion of carbon dioxide into monoxide, however, the starting point presents a more formidable problem. It is impossible to hold the constituents together by hand, and waiting for them to come together on their own takes an unpredictable amount of time, like a courtship. To overcome this difficulty, Zewail's group hit upon the expedient of letting hydrogen ride into the vicinity of carbon dioxide on the back of a carrier: they mixed CO_2 with a compound made of hydrogen and iodine. A femtosecond blast of light broke the HI apart and launched the hydrogen atom toward its carbon dioxide target. Thus the first fast laser pulse served to provide the starting signal for the experiment.

The other, unrelated function of femtosecond pulses is to illuminate the reaction after it has begun, like a stroboscope, except that the light scattered by the atoms involved is collected by spectrometers rather than photographic plates. Just as atomic spectra held the clues that enabled Bohr, and later Heisenberg and Schrödinger, to unravel the structure of atoms, molecular spectra reveal the mechanics of their chemical interactions. Each of the four compounds that participate in the carbon dioxide reaction has its own unique fingerprint in terms of the colors of light emitted and absorbed. After the reaction is initiated, the spectra of the first two compounds—the initial ingredients, hydrogen and carbon dioxide—begin to fade, and then, frame by femtosecond frame, those of the final products take over. Not only composition but speeds and relative positions of molecules can be deduced from the telltale characteristics of the spectra, which is also the way astronomers can learn so much about the stars they will never reach. But even beyond the motions of the interacting molecules, the spectra reveal something entirely new.

In the interval between the demise of the initial chemical state and the full development of the final reaction products, a new type of spectrum appears, and by means of quantum theory it was assigned to a new molecule that mediates the reaction. Described by the chemical formula HOCO, it is known as a collision complex and lives for only about five picoseconds before breaking up—much too briefly to be observed in ordinary chemical experiments, but long on the femtosecond scale. Like all complex molecules that have just undergone a violent trauma—birth, in this case—it vibrates and rotates through space. The tumbling, shivering HOCO molecule is a quantum mechanical Cheshire cat, a short-lived enigma about which little is known at present.

If traditional chemistry is an enumeration of the raw materials that are carried into a kitchen through one door, followed by a description of the gourmet meal that emerges through the other, femtochemistry allows us a glimpse inside the kitchen. It is chemistry on the move and bridges the disciplines of physics and chemistry in a way that could previously only be seen by way of the imagination, as in Kekulé's dream. Femtochemistry represents the latest link in the reductionist program begun by Leucippus and Democritus, and celebrated poetically by Lucretius, of describing nature in terms of the motion of its irreducible constituents.

When Isaac Newton found the basic laws of motion that applied to macroscopic bodies, he realized that his theory was only the beginning of a much grander scheme. In the preface to the first edition of *Principia Mathematica,* he spells out his vision:

> I wish we could derive the rest of the phenomena of Nature by the same kind of reasoning from mechanical principles, for I am induced by many reasons to suspect that they may all depend upon certain forces by which the particles of bodies, by some causes hitherto unknown, are either mutually impelled toward one another, and cohere in regular figures, or are repelled and recede from one another.

By the nineteenth century this hope had become a firm conviction, as enunciated, for example, by Hermann von Helmholtz, the discoverer of the principle of conservation of energy: "The phenomena of nature must be reduced [he used the word *zurückgeführt,* meaning "led back"] to motions of material points. . . . So at last the task of Physics resolves itself into this, to refer phenomena to inalterable attractive or repulsive forces whose intensity varies with distance." Even the names of the various branches of physics echo this ambitious but profoundly austere program. At the foundation lie *mechanics,* the study of the motion of ordinary objects, and *dynamics,* which deals with the forces that cause this motion. Motion was believed to be at the root of all phenomena. The study of heat became *thermodynamics* and *statistical mechanics,* and electrical and magnetic phenomena were summarized by *electrodynamics.* Then *quantum mechanics* transformed classical mechanics, and subsequently even the study of light, the most immaterial of materials, became *quantum electrodynamics.* When quarks were recognized as constituents of nuclei, their motions were codified by *quantum chromodynamics* (which despite its name has nothing to do with ordinary color), and at the opposite end of the scale the largest atomic particles are described by *molecular dynamics,* of which *femtochemistry* is the cutting edge. Beyond that range lie, in Newton's own words, "the operations in chemistry" and "bodies of a sensible magnitude," that is, the things around us.

"There is a straight ladder from the atom to the grain of sand, and the real mystery is the missing rung," according to Tom Stoppard. "Above it, classical physics. Below it, quantum physics. But in between, metaphysics." Molecular dynamics, and in particular femtochemistry, provides a magnifying glass that helps to locate the boundary between classical and quantum mechanics more precisely. The missing rung resides far below the grain of sand and a little above the atom, just where atoms join together to become molecules. This is the unique territory where two conflicting ways of understanding motion meet and commingle.

To account for the motions of all the particles of the HOCO

collision complex, including its four nuclei and the swarm of electrons that surround them, would be well beyond the means of existing computers, so the problem must be divided into two parts. The first is the quantum mechanical treatment of the electron cloud, in which a cluster of electrons forms a ghostly, oozing miasma. The other part is the description of the four nuclei, which are assumed to move among the electrons in a classical manner, like hailstones through a thundercloud.

Even with this simplification, the mathematical complexities of molecular dynamics are staggering. It took about forty-five years, from 1930 to 1975, to perfect a proper description of the most primitive of all chemical reactions—hydrogen atoms colliding with diatomic hydrogen molecules—and then another fifteen years to extend the analysis to simple practical examples. A complete theory of the complex reactions observed in Professor Zewail's laboratory is still a long way off. When it is finally accomplished, it will be possible to follow the motions of the nuclei like those of a falling cat, but the behavior of the electrons will remain codified by abstract lists of numbers in the bowels of the computer. Like Heisenberg's matrices, these numbers will evoke no visual images.

In some respects the science of molecular dynamics resembles the operation of a large clothing business. At headquarters a computerized inventory system follows the progress of raw materials through manufacture to the final product and thence to its distribution and sale. To determine the location of a particular red dress, for example, one simply enters the identification number of the garment and learns that it is currently hanging in a boutique in Chicago. If one wants to know its price, size, and date of manufacture, those are also instantly available.

For accountants the identification number is more real than the dress itself. Using numbers as tags, elaborate flow charts and pie graphs can be generated, and although they may be meaningless to most people, they capture the operation of the business far more faithfully than a color photo of the actual gown would. Of course an occasional trip to the field is necessary, to verify that the red frock really is where it is believed to be and that its size and color correspond to what the computer claims. To the extent that such checks are successful, the model is adequate. But a tag is not a dress, and in fact, ironically, the better the mathematical model becomes, the farther it is removed from reality. In the old days storekeepers had to close shop every now and then to examine and count their wares, but today automatic inventory control has made this step superfluous. Numbers have replaced things.

Quantum theory is the accounting system of the atomic world, a mathematical model that allows predictions to be made that can be verified in the laboratory. As calculated by supercomputers, the information contained in the wave function can be displayed in a variety of ways, such as successive stroboscopic views of the positions of the nuclei, but in all cases the electrons remain nebulous.

As we begin to chart the strange territory between classical and quantum mechanics, we learn about the limitations of both. Classical motions have the distinct advantage that intuition, formed and polished, can guide the theorist in the analysis of complex molecular systems. Quantum mechanical motion, on the other hand, is more accurate but less amendable to intuition. In the end it is the convergence of the two into a unified description that brings the richest rewards. In an article published in December 1990 Ahmed Zewail revealed the nature of these rewards when he wrote in glowing terms about the prospects of "laser customized chemistry," which will steer the motions of individual molecules and control chemical reactions at will.

In the course of describing himself in his novel *The Monkey's Wrench* Primo Levi eloquently captured the dream of total control over atoms: "I've always been a rigger-chemist, one of those who make syntheses, who build structures to order, in other words," and as an example of such a structure he draws the chemical formula for a huge molecule consisting of about seventy atoms. Then he continues:

> But we are still blind . . . and we don't have those tweezers we often dream of at night, the way a thirsty man dreams of springs, that would allow us to pick up a segment, hold it firm and straight, and

> paste it in the right direction on the segment that has already been assembled. If we had those tweezers (and it's possible that, one day, we will), we would have managed to create some lovely things. . . . But for the present we don't have those tweezers, and when you come right down to it, we're bad riggers.

While chemists want to handle atoms and strive for total control, physicists want to understand them, and though understanding is a prerequisite for achieving control, it is not necessarily understanding at the most fundamental level, which is the aim of physics. The difference between the two approaches is illustrated in a remark by Arnold Sommerfeld, Heisenberg's thesis adviser in Munich. In the first half of this century Sommerfeld, with his immaculately twirled handlebar mustache, presided over German physics like a strict but beloved father figure. His research career blossomed in the interregnum between the destruction of classical physics by Rutherford and Bohr around 1913 and the birth of quantum mechanics twelve years later, and he was hoping to develop a complete particulate theory of matter in the tradition of Democritus and Newton when the rug was pulled out from under him by his star pupil's radical revision of atomic theory. To his credit Sommerfeld was immediately and fully converted, and became a vigorous proponent of quantum mechanics.

In May 1928, Sommerfeld was invited to give the keynote address at a meeting of the German Society for Applied Physical Chemistry in Munich. The title of his address was "On the Question of the Meaning of Atomic Models." Feeling a bit like a philosophical Daniel in a den of eminently practical lions, he began with "The foolhardy idea to represent an atom by a model," and then he explained his choice of words: "I call the idea of an atomic model foolhardy. So it seemed to the majority of physicists and chemists until long past the year 1900. Imagine, for example, how our venerable and stern Adolf von Baeyer would have treated someone who tried to explain to him the advantages of an atomic model."

At the University of Munich, where my great-grandfather

had been professor of chemistry for forty-two years, his autocratic manner was legendary, and his colleagues also knew that as a successful rigger of molecules he had had no use for speculation about the ultimate nature of matter.

I suspect that modern femtochemists have as little patience with metaphysical questions about the reality of atoms as my great-grandfather would have had with Sommerfeld's quest for intuitively appealing atomic models. On the other hand, a different, more practical extension of technique that physicists are developing is of considerable interest to chemists—the perfection of tweezers for manipulating single atoms. But before one can watch the interactions of individual atoms in time and space, one must first learn to locate and count them one by one.

9

Counting the Atoms

The town of Oak Ridge, Tennessee, was built by the atom. Two generations ago the site was an isolated forest in a valley between the Great Smoky Mountains and the Cumberland range. Then, in 1942, it was selected as the headquarters for the Manhattan Project, the American effort to build an atomic bomb. A security fence was erected, and a boomtown with 75,000 inhabitants, workers at the secret fissionable materials plant, mushroomed behind it. When the first atom bomb exploded in the desert of New Mexico in the summer of 1945, the town's purpose was achieved.

Soon after the war the fence came down, and the shopping centers, housing developments, fast-food restaurants, and four-lane highways of contemporary America began to move in. Today the town bears little resemblance to its wartime appearance, but its economy is still tied to the atom. The bomb factory continued production throughout the cold war, and its research division, the Oak Ridge National Laboratory, became a world leader in the development of peaceful uses of atomic energy. The words *nuclear* and *radiation* inspire no terror in the people of Oak Ridge, because most of them work in industries that produce, besides ingredients for weapons, radioactive pharmaceuticals, instruments for monitoring the environment, safety equipment for nuclear power plants, and a host of other scientific, environmental, and medical applications of nuclear technology. Oak Ridge is the site of an atomic energy museum, and of the headquarters of a consortium devoted to nuclear research that comprises sixty-three American universities. For physicists the name Oak Ridge represents a welcome counterpoint to the horrors evoked by the words Hiroshima and Chernobyl.

I went to Oak Ridge to meet Sam Hurst, an atomic entrepreneur who is known as the inventor of a reliable and versatile method of counting atoms that has applications in the most diverse fields of science and technology. In addition to learning about his technique, I wanted to experience an atmosphere in which atoms are ordinary, everyday objects, like cups and saucers and grains of sand. Ahmed Zewail's stroboscope, Daniel Kleppner's vacuum cavity, Hans Dehmelt's atomic trap, and the scanning tunneling microscope of Gerd Binnig and Heinrich Rohrer are all important instruments for the manipulation of atoms, but they are still only found in the exotic surroundings of academic research laboratories. Practical scientists such as Sam Hurst, on the other hand, have taken their atomic tools out of the laboratory and into the doctor's office, the factory, and the kitchen; the atoms themselves they take for granted. In Ernest Rutherford's words, they can almost see the jolly little beggars. In Oak Ridge, atoms are more manifest than elsewhere in the world.

Hurst met me at Atom Sciences, Inc., the little business he founded in order to commercialize his ideas. It is located in the Ridgeway shopping center, past a supermarket, next to a contractor's office and a store enigmatically called Family Tailor & Gift Store. Sam has retired from the business, but judging by the attitude of the receptionist in the tiny lobby and the return address on his letters, I gathered that he is still as active as ever. A native of Pineville, Kentucky, he is short and trim—in a commemorative photo of a visit by Jimmy Carter the two men resemble each other in stature, but Hurst's mien is more serious than the president's; with reading glasses his somber look gives him a somewhat owlish appearance. His sentences are delivered with a mountain twang, and they come slowly—each one chosen with deliberation for accuracy and succinctness and interspersed with thoughtful pauses. He comes across as a beguiling combination of internationally renowned scientist, homespun philosopher, and country boy who made good.

Hurst's career unfolded in four different arenas of scientific research: the government laboratory, the academy, the private sector, and the think tank. After developing the universal atom counting method at Oak Ridge National Laboratory, he joined the faculty of the University of Tennessee in nearby Knoxville, founded Atom Sciences, Inc., and helped to create the university's research institute devoted to the extension and worldwide promulgation of the technique.

As he showed me around the little empire he has built, Hurst patiently told me the story of atom counting. We wandered through a warren of little rooms, each one dedicated to a separate operation, in which half a dozen young technicians sat behind computer consoles or fiddled with vacuum pumps and dye lasers, and Hurst had a joking greeting or technical remark for most of them. In every available corner there were samples of materials that had been sent here for analysis: a canister full of air extracted from bubbles in the ice of a glacier in Alberta, a bottle of water from Tunisia, a box of semiconductor chips from Russia; the place was an international smorgasbord of atoms.

Like many inventions, the universal method of counting atoms was born of necessity. In 1970, as leader of a group at the Oak Ridge National Laboratory charged with investigating certain nuclear reactions, Sam Hurst encountered a problem. He believed that an anomaly in his data resulted from the presence of impurities in his materials, but the levels of contamination were so small as to be undetectable by any known method. Chemists routinely deal with quantities in ppm—parts per million—but in view of the vast number of atoms in a drop of water, a ppm is not really a small amount. Hurst believed that the impurities he was dealing with were well below the ppm level and decided to figure out how to detect them.

Knowing that modern electronics can reliably count small numbers of electrons, down to a single particle, he reasoned that if one electron is stripped from the outer shell of each atom and then the electrons are counted, the number of atoms will be known. Furthermore, lasers can be used as versatile and efficient guns for knocking electrons out of atoms, only the method is not in the least bit selective and strikes all types of atoms indiscriminately. A sufficiently powerful laser will remove electrons from every atom it illuminates, which, of course, is counterproductive; the point is to detect only atoms of a specific kind in order to measure minute impurities in a material.

Then Hurst had an inspiration. Suppose, he thought, the laser is adjusted so that it doesn't knock the electron all the way out of the atom but only a little more than halfway up the energy staircase. Since the sizes of the energy steps are different for each atom, and modern lasers finely tunable in energy (or, equivalently, in frequency, wavelength, and color), the excitation of selected atoms can be achieved with exquisite discrimination. Therefore, a laser adjusted to raise an electron in, say, an aluminum atom will have no effect at all on a nearby atom of oxygen.

The last step then follows easily. The laser delivers such enormous numbers of photons with each burst that after all the atoms of a certain type have been excited, there will still be plenty of photons left over. Every atom that has had its energy increased will then absorb a second photon of the same kind, which will knock the electron the rest of the way up the energy staircase and over the top, and thus completely out of the atom, where it is finally counted.

The crux of the method lies in the first step, which represents a resonance between the laser light and a specific type of atom. Tuning a laser to excite only selected atoms is very much like tuning a radio by establishing a resonance between the receiver and the station's emitter. Just as a good radio can discriminate between a multitude of adjacent stations, a good laser can select among a multitude of different atoms. When Hurst and his team realized, in 1974, that with some variations the method could be adapted to count concentrations down to single atoms of any element, they applied for a patent, which was granted two years later. Additional patents on various refinements followed. By 1975 Hurst and his colleagues succeeded in detecting a single cesium atom in a background of 10^{19}, or ten billion billion, argon atoms. Today Hurst, and Atom Sciences, Inc., of Oak Ridge are considered to be the reigning experts in the counting business.

The success of the new invention in finding a few atoms among trillions of others began to attract attention, and Hurst's method was soon adopted by laboratories throughout the world to tackle a great variety of practical problems. One satisfied customer is the computer industry. As integrated circuits on computer chips become ever smaller, the likelihood that minute faults in the material will spoil their delicate electrical performance becomes increasingly serious. Some modern electronic components are so small that a single foreign atom can cause them to malfunction. Known as single-atom failure, this problem may one day be a common nuisance in everyday life and will be prevented only by heroic new methods of chemical analysis, of which Hurst's resonance technique is a precursor.

Basically, the problem is one of geometry. A single atom resembles a point, which mathematicians call zero-dimensional. By the same token a stream of atoms, like a stream of water from a hose, looks like a line, and is one-dimensional, while a surface examined by scanning tunneling microscope is a two-dimensional structure. But a bulk sample in which an impurity is to be detected is intrinsically three-dimensional. With each dimension the number of atoms, and hence the difficulty of sorting, increases dramatically.

Consider a cube of metal with a width of one micron, a millionth of a meter, the size of certain current commercial microchip structures. There are about ten thousand atoms along each edge, a number that is large, yet manageable. On each surface the number of atoms is ten thousand squared, or a hundred million—a more formidable quantity. But the entire block consists of ten thousand cubed, or a trillion, atoms. Finding one atom in a trillion is the kind of challenge the atom counters are up against. The problem is not so much one of handling single atoms as it is of disregarding myriads of others. The watchwords are discrimination, selectivity, sensitivity, and specificity, rather than magnification, which matters in microscopy, and stability, which is essential for trapping. In his search for a needle in a haystack, Hurst has discarded the conventional approach of sorting through the blades of hay and brought in a magnet, which ignores them.

The atoms found by Hurst's method are not always as unwelcome as impurities. Consider this imaginary but plausible scene: In a secluded valley of northern China two bright-faced young men clad in colorful Western hiking clothes, each carrying a small nylon backpack, scramble down a slope to the bank of a bubbling mountain stream. Pushing his sunglasses into his black hair, the first one kneels to scoop up some sand in a little vial, while the other one records an identifying number in his notebook. They confer briefly, stow away their equipment, and set off again upstream. Although their sporty appearance belies it, they are prospectors, and they are engaged in the world's most sophisticated method of panning for gold.

Later, in a hamlet farther down the valley, in a panel truck converted into a portable laboratory, they test the sand for the presence of this precious metal. But they are not looking for whole nuggets, or even flecks—they are counting single atoms. The reasoning behind this approach is that although gold is impervious to corrosion and most normal chemical interactions, it is not indestructible on the atomic level. If there is a clump of gold somewhere upstream, individual atoms are continuously worn off it by the incessant action of water and the surrounding stones and sand. Billions of atoms are removed in this way, and below the clump a plume of gold atoms usually spreads out for miles downstream. The high-tech prospectors map out the extent of this plume by measuring concentrations of gold that are far too puny to be of any use whatever. Since traces of gold are present everywhere on earth, the actual density of the plume does not tell them much either. Changes in the concentration of gold atoms are far more informative: Increases point the way in the direction of useful deposits, decreases away from them.

After learning about the technique from Hurst when he visited them several years ago, a group of physicists at Tsinghua University in Beijing developed a spectacularly sensitive analytical system for the detection of gold and other metals in minerals. In their published papers they report their findings in terms of the abbreviation ppt, for parts per trillion—a pretty discouraging amount for a conventional gold hunter. However, they have succeeded in demonstrating the feasibility of their novel technique and will soon be able to put it to use to enhance the urgently needed economic development of their country.

In the United States, where gold mining is lower on the national agenda, atom counting has been applied to other societal problems. One intriguing possibility is the beginning of a technique called single-atom medicine. It has been known for many decades that when the nucleus of the boron atom is hit by a neutron from a nuclear reactor, it emits an alpha particle—the projectile used by Ernest Rutherford in his historic experiment. If this alpha particle were released within living tissue, it would travel no farther than the edge of the cell it happened to inhabit before it expended its energy and came to a stop. Then it would attract two stray electrons and turn into a harmless inert helium atom. But the energy it imparted to the cell in the process of slowing down would kill the cell, making neutron irradiation of boron atoms a promising candidate for cancer therapy.

The idea is appealing. Instead of drowning a cancer in a flood of noxious chemicals or blasting it indiscriminately with a volley of powerful radiation, a doctor could pick off the disease, cell by cell, without harming the rest of the body. This kind of microscopic control over organic processes is the dream of the medicine of the future.

At the Idaho National Engineering Laboratory in Idaho Falls, there are plans to adapt a nuclear reactor for the development of this promising therapeutic technique. But before the first patient can benefit from it, much remains to be done. For example, a reliable method of introducing boron atoms into the cancer cells must be perfected. It is not necessary to restrict the boron atoms to one per cell, but the number must be small enough to prevent an excessive amount of energy from being released into a single cancer cell. Thus the movement of minuscule numbers of boron atoms through the human body must be closely monitored and controlled. When Sam Hurst heard about the problem, he knew that his counting technique had found yet another significant application.

Other kinds of medical uses of atom counting include the study of the effects of extremely small traces of various elements on the human body, and the potential for reducing the sizes of the samples required for various laboratory tests. Both of these applications combine in a recent discovery in neonatology. In view of the size and delicacy of a newborn baby, blood samples must obviously be kept at an absolutely minimal size—but small concentrations of elements in minuscule samples are undetectable by conventional techniques. For this reason Atom Sciences, Inc., has been collaborating with a team of pediatricians and a chemical laboratory in Maryland to examine the problem. What they have found is that the traces of metals such as chromium, iron, copper, and nickel that are required for normal development are transferred from the mother to the fetus rather late in pregnancy, and that very premature babies may therefore lack these elements and suffer various ailments and birth defects. The researchers therefore aim to learn how to supply the proper amounts of necessary trace elements to premature babies. The amount of blood required for this purpose is small, measured in minute drops, but still represents boundless oceans on the atomic scale.

Yet another application of atom counting, as significant as safeguarding the purity of electronic materials and following the movement of elements through the human body, is found in the field of ecology. Even as our understanding of the atmosphere and oceans grows more global, it is also becoming increasingly microscopic. The large and the small meet not only in the realm of cosmology, where quarks and leptons constitute the stuff of the Big Bang, but here at home, too, where the details of atomic interactions ultimately determine the future of the planet.

Atom Sciences, Inc., has become instrumental in unraveling the history of the water reservoirs of the world. In principle the method is simple and resembles the time-honored technique of carbon dating: If you know how many radioactive atoms of some type were present in a given sample of water at some specific starting time and you know how fast they decay you can then measure how many of them remain and deduce the time that has elapsed since the initial moment. Certain atomic species, such as krypton-81, are continuously replenished in the atmosphere by the action of cosmic rays, so the concentration of krypton-81 in air at sea level is assumed to be a constant that has not varied for millions of years. It is furthermore known that as long as water is in contact with the atmosphere, it contains a fixed, minute concentration of krypton-81.

Suppose that at some time long ago a quantity of water was somehow sequestered and protected from the atmosphere, by trickling into an underground reservoir, for example, or by becoming fixed in ice. At that moment the number of krypton-81 atoms in the sample begins to diminish at a steady pace, and their concentration, measured today, serves as a clock to determine the time of sequestration. The numbers, at both the beginning and the end of the process, are small: A quart of water contains a thousand krypton-81 atoms in the beginning and half that number after two hundred thousand years. When you deal with concentrations at that level, you are not engaging in chemical analysis—you are counting atoms.

In this way the age of groundwater and polar ice—which is to say, the elapsed time since they were last exposed to the atmosphere—has been measured. Such information is necessary for understanding the history of the surface of the earth, and its trends, such as global warming. For processes that unfold more quickly, other trace elements besides krypton-81 are more appropriate. For example, argon-39, with a half-life of 270 years, is used for monitoring how quickly ocean water rises from the depths and then descends again, and how long ago water accumulated under the Sahara desert.

The list of present and future applications of Sam Hurst's invention is endless, but beyond that the technique has a subtle philosophical implication. Chemistry has portrayed a world clearly divided into different substances—the ink on this page is primarily carbon, the air we breathe consists of oxygen and nitrogen, my wedding ring is made of gold. Although chemists have always understood that all substances are riddled with impurities, most of the minor constituents were too rare to be measured and could be safely ignored. Consequently every substance was labeled, at least in principle, like a candy bar, with a short list of its most significant ingredients and a brief appendix of additional trace components at the edge of detectability.

But since 1970, when Hurst first envisioned his scheme, this way of comprehending the world has changed. The limit of detectability for every element has been reduced to its theoretical minimum, the single atom. Today there is no longer such thing as a concentration that is too small to be measured: either an element is present or it isn't. A chemical assay is now different from the biological survey of a patch of ground, which must inevitably end with a phrase such as "plus an innumerable quantity of invisible microorganisms." The chemical composition of matter is now, in principle, absolutely known.

With such a dramatic increase in the sensitivity of analytical chemistry, most naturally occurring substances must be considered to contain atoms of every element. A grain of sand, for example, which was once thought of as consisting almost entirely of quartz, a compound of silicon and oxygen, with a minute admixture of trace elements, has been found to contain a few atoms representing practically the entire table of elements. The qualitative classification of the world into different substances now becomes a quantitative classification in which the period at the end of this sentence is no different in kind from the ring on my finger—both contain carbon and gold atoms, only the proportions differ.

Another consequence of this improved analytical technique is the completion of the inventory of the world begun by Democritus. As long as many of the ingredients of ordinary things were too scarce to be identified by any known method, the notion of a universe composed of atoms was still an abstraction. But today it is possible, in principle, to identify and list every single atom in any given object. What is the world made of? Atoms, answers Democritus, and Atom Sciences, Inc., can tell you which ones.

At the end of my visit, after a long day of explanations, introductions, and demonstrations, Sam Hurst took me to his secluded hilltop home in the forest that grew here long before the atom put Oak Ridge on the map. As we wound down our conversation over a beer, he showed me one of the thick notebooks of overhead transparencies he takes along on the trips he makes to introduce his method to colleagues throughout the world. To my astonishment it was filled with page after page of poetry:

> . . . water gives way to the fish
> As it swims, and opens a passage for it to pass,
> Because there is a space left behind the fish
> Into which the liquid can flow: and this, they say, demonstrates
> How other things can change place, although space is full.

and so on, for a hundred pages or more. It was Lucretius. Hurst, it turned out, is a serious fan of the Roman poet and has culled an ingenious collection of quotations to illustrate every conceivable technical point that can come up in a discussion of atom counting. If Lucretius was a prophet, Sam Hurst is clearly his apostle.

I was delighted to know that Hurst places his own work into its rich historical context, all the way back to the beginning of the Common Era. The fascination of Lucretius is not only that he had anticipated so many details of modern science but that through his poetry he tried passionately to bring science into the lives of his contemporaries, no matter how indifferent they seemed to be to it, and perhaps even to learning of any kind. Sam Hurst realizes how desperately urgent this same concern is in the contemporary world, and I applauded his identification with Lucretius as proselytizer of the atomic doctrine.

At the same time I was troubled by this preoccupation with the Roman poet, almost to the exclusion of everyone who came later. The theory of Leucippus and Democritus that Lucretius taught is only the opening scene of the story of the modern atom. The world consists of atoms, yes, but what are they? Why was Sam Hurst not equally fascinated by the ascetic Werner Heisenberg, who believed that we cannot fashion an intuitively appealing image of the interior of the atom, or the single-minded Erwin Schrödinger, who wanted to reduce all material phenomena to waves, or the radical Max Born, who interpreted atomic reality in terms of probabilities? Perhaps Hurst was wise to stop at a point of certainty, the fact that we are made of atoms, without venturing out into the swamps of incomprehensibility and doubt that beset those who venture into the meaning of quantum mechanics. But I was worried that by choosing the zealous Lucretius as his hero, he might succeed all too well in communicating to educated laymen the exhilarating experiences of his own discoveries and therefore fail to acknowledge that the most important questions remain unanswered.

Throughout his career Hurst has relied on quantum mechanics to describe the creation of laser light, and its interactions with atoms, so his experience of both the power and the limitations of the theory is extensive. In this respect he is typical of the majority of modern physicists, to whom quantum mechanics at the workaday level is no more mysterious than auto mechanics. The paradoxes of its interpretation do not particularly concern him; what matters to him is that he knows the rules and that they work.

During my visit Hurst strayed from this pragmatic position only once. At lunch in a fine seafood restaurant in a valley out of view of the sprawling modern town of Oak Ridge, we talked about recent experiments on the interference of particles. When the conversation turned to the subject of Young's double-slit experiment performed with a beam so feeble that only one particle passes through the apparatus at a time, Hurst mumbled, "I just don't understand that!" and then, "That one really bothers me." Even thorough familiarity with the nature of atoms cannot dispel their ineffable aura of mystery.

For the twenty-first century Lucretius will not suffice as prophet. A new generation of physicists, building upon the foundations laid down by the architects of the quantum theory, will have to lead us beyond the mere recognition of the existence of atoms, to penetrate their forbidding cores.

The Future

10

Atomic Standards

To explore an atom, you have to measure it. Unless you quantify its properties, such as size, weight, and characteristic frequencies, you cannot understand its nature. Measurement, however, is really nothing but a comparison with accepted standards set up for that specific purpose. So before you can tame an atom, you have to choose the standards. But how can you measure the invisible? What sets the standards in a world at the edge of perception?

As long as the human being was firmly enthroned as the central concern of philosophical inquiry, it was natural that the standards of measurement, too, were human. The Greek philosopher Protagoras, a native of Abdera like his younger colleague Democritus, put the matter succinctly: "Man is the measure of all things." For the next two and a half millennia physical standards were chosen according to this criterion. The legal definition of the foot, for example, was described as follows in a book on measurements published in Germany in 1522, two generations before the dawn of modern science: "Stand at the door of a church on a Sunday and bid sixteen men to stop, tall ones and short ones, as they happen to pass out when the service is finished; then make them put their left feet one behind the other, and the length thus obtained shall be a right and lawful rood to measure and survey the land with, and the sixteenth part of it shall be a right and lawful foot." Though this picturesque prescription is easily understood, and even has a ring of undeniable authority, it is hardly a sufficiently precise standard for surveying the atom.

In physics time, length, and mass form the basic triad of dimensions used for measuring things and processes, including atomic frequencies, sizes, and weights. Time itself is as difficult to define as it is intuitively obvious. In A.D. 400, Saint Augustine summarized the problem by writing, "What is time? If no one ask of me, I know; if I wish to explain to him who asks, I know not." But the *units* of time—seconds, minutes, hours, days—are simple and accepted by the entire world.

Ironically such universal agreement is lacking in the measurement of length, which is a much more concrete concept than time. In the United States feet and inches are still in common use, even though scientists, as well as most of the rest of the world, rely on the metric system. To simplify matters, the inch has lost its independent status and is defined as 2.54 centimeters, exactly. By tying the two units together mathematically, those nations that have difficulty switching from an archaic system of measurement are at least relieved of the inconvenience of maintaining a separate official prototype ruler to serve as the standard of length.

Mass is closely related to weight, but it is a more useful measure of the quantity of matter in an object because it is independent of location. The mass of a kilogram of sugar is the same in New York, on the moon, and in outer space, where its weight would amount to two and a quarter pounds, a third of a pound, and zero, respectively. Mass measures inertia, the tendency of material objects to resist the influence of external forces, and this resistance does not depend on gravity or location: Even in weightless outer space a spaceship is much harder to move around than a screwdriver, because its mass is so much greater. Although the definition of mass is intuitively straightforward, its measurement is as problematic as the determination of time and length. As with length, there are several different units of mass, such as the preposterously named slug, but they, too, are defined in terms of a universal metric measure—the kilogram.

To facilitate fair international trade, and the comparison of scientific measurements from laboratories throughout the world, it was agreed in 1875 that the second, the meter, and the kilogram would be the international standard units of measure. In that year the government of France, in the spirit of reform that accompanied the founding of the Third Republic, decided to capitalize upon the leadership of French scientists in developing the metric system and called an international conference on measuring standards in Paris. The document that was hammered out by the participants, called the Treaty of the Meter, was ceremoniously signed on the twentieth of May 1875 by diplomatic envoys from seventeen nations, including the United States, and established the International Bureau of Weights and Measures, with headquarters in the Parisian suburb of Sèvres, as the world's keeper of standards.

The offices of the bureau are located in the Pavillon de Breteuil, a charming little eighteenth-century château in a wooded park next to the famous porcelain factory of Sèvres. The symmetrical perfection of the Pavillon and its immaculate grounds conveys a sense of serenity befitting its mission of guarding an international treaty. The quiet elegance of the compound, which enjoys the extraterritorial status of a diplomatic residency, contrasts with the robust, functional appearance of the National Institute of Standards and Technology in Boulder, where the emphasis is less on preserving old standards than on inventing new ones.

The actual prototype meter, a sturdy rod made of platinum and iridium alloy, and the kilogram, a fist-sized cylinder made of the same metal, remain locked in a vault under the Pavillon de Breteuil. According to the Treaty of the Meter, the strong room can only be opened by the simultaneous use of three different keys entrusted to three officers of the bureau. Inside, the International Meter lies in its black protective casing, which resembles a large mailing tube, on the upper shelf of a heavy steel box. On the lower shelf, amid instruments for monitoring the temperature and humidity of the surrounding air, the International Kilogram sits on a quartz base and is covered by three protective bell jars enclosing each other like Russian dolls. It is flanked by six secondary standards, exact copies of the first one, in smaller bell jars. To the few scientists who are privileged to see it in person that box must seem like some sacred shrine.

Both the International Meter and the International Kilogram were put in place in 1889. (Since a common definition of the second already existed, a new standard was not deemed necessary.) Since then, whenever a London bartender called the time, a Paris couturier measured out a half meter of silk, or a Ukrainian peasant woman haggled over a kilo of potatoes, they referred, however indirectly, to these standards. The second hand of a clock proceeded according to an astronomical cycle, each step equal to 1/60 of 1/60 of 1/24 of a day measured from noon to noon, when the sun reaches its highest point in the sky. The wooden yardstick and the grocer's scale were calibrated, crudely to be sure, against the ultimate prototypes of length and mass in that vault in Sèvres.

In the face of wars and violent revolutions, against a background of broken agreements and empty promises, in a world ruled by passion and discord more often than reason and harmony, the Treaty of the Meter has stood firm, kept alive and vigorous by the international community of scientists whose common vision transcends national boundaries and political differences. The prototype standards represent the scepter and the orb of the great empire of international industry and commerce, symbols of order and stability as well as rare reminders that in human affairs rationality and cooperation can prevail.

In practice the complex trail that leads from the common measurements that we make each day to the genuine international standards is almost impossible to follow. The manufacturers of watches, rulers, and scales rely on factory standards that in turn are calibrated against more accurate provincial and national standards made under the supervision of government agencies charged with ferreting out error and fraud. Thus the calibration proceeds by successively more precise and elaborate stages.

All along the way outside influences must be eliminated or at least compensated. Pendulum clocks vary with altitude above sea level, meter sticks expand with heat, and scales register vibrations caused by passing trucks. Such phenomena influence all measurements, and it is the business of professional metrologists like David Wineland and his colleagues in Boulder to sort out their effects and ensure that seven meters in Tokyo equal seven meters in New York.

Despite technical complications, in principle the process of measurement used to be fairly straightforward: To find your height, place a standard meter alongside your body and read off the result. To weigh beans, heap them into one pan of a balance scale and place a standard kilogram in the other. To determine your pulse rate, count your heartbeat while watching the second hand of a watch. For the first half of this century metrology consisted of developing refined versions of these procedures.

But science marched on. As measuring instruments became ever more sensitive and theoretical and experimental work reached unimaginable levels of precision, the old standards grew inadequate. For the sake of accuracy it became necessary to refine the basic units, a process that has introduced exotic new standards and made the science of metrology less and less comprehensible to nonscientists. At the current rate of progress metrology will soon be cut off from its ancient roots in commerce and daily life.

One of the first units to leave the realm of common knowledge was the second. For millennia the most accurate clock had been the earth itself. By gauging its rotation with reference to the sun, the day was defined, and then by division of the day into equal parts—twenty-four hours, sixty minutes, and sixty seconds—the second was determined. But atoms keep time too; the light and radio waves they emit have definite, unvarying frequencies, or oscillations per unit of time. By the 1950s practical atomic clocks based on this constancy had been developed.

The modern atomic clock represents the culmination of the efforts of many people, but the undisputed leader in the field is Norman Ramsey, of Harvard University. A tall, courtly gentleman with oversized round glasses and a ready smile, Ramsey is known in the profession as one of the most accessible of the great physicists. He answers his own telephone and never seems to be too busy to enliven his conversation with an anecdote or quip. About his own work he says that he got into it while searching for a dissertation topic at Columbia University in 1937; at that time he was told by his professor that there was little future in the study of atomic beams. Being of an independent mind, he ignored that advice, and fifty-two years later proved it wrong in a definitive way.

In 1989 Ramsey received half of the Nobel Prize for his work on atomic beams and his contributions to the development of the atomic clock. One of his innovations was the hydrogen maser, a device for producing precise radio waves the way a laser produces pure light, which he invented together with Daniel Kleppner of MIT. The precision of a hydrogen maser made possible the exquisitely delicate tracking of the Voyager spacecraft that enthralled the world with its color photographs of the solar system. The other half of the Nobel Prize, incidentally, was shared by the pioneer atom trappers Hans Dehmelt and Wolfgang Paul. The intimate connection between the contributions of the three laureates is embodied in David Wineland, Dehmelt's former assistant, who is trying to use a Paul trap to improve Ramsey's atomic clock.

The heart of an atomic clock is an atomic beam. The clock numbered NBS-6 in Boulder, for example, consists of a gleaming stainless-steel vacuum tube, about ten centimeters in diameter and three meters in length, mounted on a strong laboratory bench. At one end a small box functions as an electric oven to heat a piece of shiny cesium metal, the same element Kleppner used in his experiment to demonstrate the malleability of the vacuum. At the temperature of boiling water cesium is still firmly solid, but a few atoms evaporate from its surface and fly off down the tube. Some of the speeding atoms are in their lowest possible state of energy, while others have a slightly higher energy because their outermost electron has been bumped into a higher orbit. Both states of the atom behave like minuscule magnets, but with different strengths. A powerful external magnet near the entrance of the vacuum tube exploits this difference and deflects the more energetic atoms out of the way, leaving only a pure beam of low-energy cesium atoms to continue down the pipe.

The middle of the tube is the actual timekeeper. A radio transmitter, tuned as precisely as possible to the microwave frequency of 9,192,631,770 oscillations per second, fills the tube with photons. Their energy, derived from their frequency by means of Planck's constant, is the precise amount required to bump a cesium electron into its higher state of energy; photons whose frequency departs from the correct one by even a hair's breadth do not match the appropriate step size of the atom's energy staircase and therefore speed past the atom without affecting it at all.

At the far end of the beam pipe, a magnet, identical to the first one, again sorts out the higher energy atoms, but instead of shunting them aside it sends them to an electronic counter, which monitors their arrival. Whenever this counter senses that the number of arriving atoms drops a little, because the radio transmitter has drifted from its preassigned microwave frequency, that frequency is automatically corrected until the count rises back to its maximum level. By this feedback mechanism the atoms keep the radio transmitter permanently tuned to a fixed, unvarying frequency. The radio, in turn, functions like the quartz crystal in a wristwatch to regulate a clock that is accurate to within one second in three thousand years.

The next generation of atomic clocks, which will improve on this accuracy by a factor of a thousand or more, is currently under development at the Time and Frequency Division in Boulder, as well as in other laboratories around the world. David Wineland explained to me that the erratic motion of the atoms down the beam pipe of a conventional atomic clock causes small but unavoidable shifts in the frequency of the radio waves they absorb. These variations are akin to the sharp drop in the pitch of a car's tire noises as it passes a stationary observer. An approaching source tends to squash the emitted sound waves together like a concertina, which raises the perceived pitch above normal, whereas a receding source draws the sound out to a lower-than-normal note. The frequency of a radio wave is analogous to the pitch of a sound wave, so the inevitable slight variations in the speed of cesium atoms, both along the tube and off to the side, alter their frequencies in an uncontrollable manner. A clock built around a single atom at rest in a trap avoids this difficulty and is therefore potentially much more accurate. Nevertheless David believes that many years will elapse before such a device becomes a practical reality.

As soon as the first atomic clock had become sufficiently reliable, it was pressed into the service of metrology. In 1967 the General Conference of Weights and Measures, meeting for the thirteenth time at Sèvres, redefined the second as the duration of 9,192,631,770 cycles of the radiation corresponding to the transition between the two lowest energy levels of cesium. The second was transformed from an astronomical unit into an atomic one, and its definition removed from the grasp of schoolchildren. Not even physicists can be expected to memorize the string of digits that characterizes it, and most of them would be hard pressed to describe the inner workings of an atomic clock.

Compared with the new second, the rotation of the earth, once thought to be as steady as the beat of a pendulum, turned out to be highly irregular. The causes of this phenomenon were soon found: Great masses of material, such as polar ice and currents of moist air, moving closer to the poles and thus to the earth's axis, speed up the planet's rotation in the same way that a figure skater accelerates her spin by pulling her arms closer to her body. Modern cesium clocks are so accurate that they can detect even the infinitesimal shortening of the day that would occur if all Americans simultaneously drove their cars five hundred miles to the north. Measured by an atomic clock, the earth's rotation varies seasonally and, in addition, displays an extremely small, but nonetheless real, slowing trend. This fascinating effect, which is the subject of current research, requires a leap second, a tiny interval of timelessness, to be inserted occasionally between the end of one year and the beginning of the next. In this way metrologists manage to keep their precise clocks in sync with the irregular heartbeat of planet Earth.

Though the new meaning assigned to the second removed it from common experience, the shift can be thought of as an improvement on an old idea. The change in the definition of the meter is much more radical. On October 20, 1983, the meter was redefined by international agreement as the distance that light travels through a vacuum in 1/299,792,458 of a second. Thus, the meter, which used to be an independent unit, is now derived from the standard unit of time. Any change in the experimental determination of the duration of a second automatically entails a change in the meter. Furthermore, the speed of light, which has been measured by astronomers and laboratory physicists with increasing accuracy for centuries, has ceased to be an experimentally determined quantity. The new definition of the meter implies that it is exactly 299,792,458 meters per second, not as a matter of observation but as a mathematical certainty. It is a ratio of units, like the ratio between the lengths of an inch and a centimeter. When one uses the modern definitions of the meter and the second, measuring the speed of light is as pointless as counting pennies in a dollar.

The most revolutionary outcome of the meter's new definition has been its effect on the process of measurement. From time immemorial, distances were determined with rulers. Now they are measured with clocks. The correct way to determine your height is to let a pulse of laser light travel from your head to your heels, to record its transit time with an atomic clock, and to convert that time, by means of the definition of the meter, into a distance. Thus, indirectly, the meter has joined the second in its journey from the macroscopic to the atomic realm.

Other units have also begun the march toward atomic standards. On the first day of 1990 the volt and the ohm, measures of electrical potential and resistance respectively, were redefined in atomic terms. The Frankensteinian electrical devices that had served as electrical standards since the nineteenth century—vats of acid trailing coils of copper wire for defining the volt, long rigid platinum rods suspended from springs attached to room-sized scaffolds for fixing the ampere—have been abandoned.

Electricity is now gauged by the quantum mechanical dance of electrons in cunningly constructed solid-state devices smaller than peas.

Ironically, even though the new volt and ohm are far more accurate than the old ones, they have not yet been officially adopted as legal, fundamental units. The international-standards business is profoundly conservative. A certain rigidity is the price for stability, in human institutions as much as in large buildings, and any agreement that has withstood the test of time as successfully as the metric system should only be amended after the most thorough consideration of all possible arguments by all partners.

The fundamental quantities of physical science are tied in a Gordian knot of such intricacy that to change one unit entails changes in many others. The dependence of the meter on the second is a simple case in point, and other examples abound. Units from seemingly unrelated disciplines are intertwined: A volt squared, divided by an ohm, is a watt, and that unit of mechanical power can in turn be reexpressed in terms of seconds, meters, and kilograms. The new electrical units cannot, therefore, be adopted as legal units along with the second, the meter, and the kilogram: to define the watt in two different ways would be to enact a logical inconsistency. At present the new volt and ohm are called "practical" units, to distinguish them from basic ones, and serve as useful interim conceptions until the international scientific community is prepared to revise the entire edifice of standards.

Of the three basic mechanical units, only the kilogram remains tied to its original prototype, the International Kilogram of 1889. That the exquisitely precise measurements of the masses of the electron and other elementary particles, not to mention every pound of flour bought and sold around the world, should still be based on a little metal cylinder in a French vault seems quaint, inconvenient, and almost incredible in this high-tech age.

The new definitions of the second and the meter allow these units to be established by any reasonably well-equipped laboratory, without reference to a central bureau of standards. But to calibrate a mass against the primary standard, the sample still has to be transported to the vault in Sèvres—a risky, expensive, and cumbersome procedure. The American standard, for example, which was the twentieth of forty nearly identical weights made by the London firm of Johnson, Matthey, and Company in 1884, has been taken to Paris only four times. Most recently, in 1984, it was removed from its protective container in Washington, D.C., by means of tongs equipped with clean, padded clamps, placed in a specially constructed traveling case, and carried in the passenger compartment of a transatlantic plane. Along the way it was handled by two attendants—one to hold it and the other to catch it if his partner stumbled.

Why can't the kilogram be defined in atomic terms too? Since all atoms of the same species are assumed to have exactly the same mass, it would make sense to use one of them as a standard. Why not let the carbon-12 atom, say, of which there are untold numbers throughout the universe, be the prototype, instead of the platinum cylinder in Paris?

The suggestion that all standards of measurement, not just mass, should be atomic, was first made by the Scottish physicist James Clerk Maxwell in 1870. Maxwell's idea had its origin in the ancient doctrine, adumbrated by Melissus of Samos, that atoms are truly identical, while all macroscopic objects and artifacts are necessarily different when measured on the atomic scale. But this program could not be implemented until a century later, when scientists finally learned to manipulate atoms one by one. Now that atoms are beginning to enter the world of everyday objects, they suddenly become the natural replacements for the imperfect, perishable artifacts that have served as prototypes for so long.

The practical obstacle to the adoption of an atomic standard of mass lies in counting. For example, if a jeweler wished to know the mass of a gold ring, he would have to know how much a gold atom weighs relative to a carbon atom and also how many gold atoms were in the ring. The first quantity, the relative masses of two atoms, can be accurately determined by instruments such as mass spectrometers, which operate by comparing the curvatures of the paths of different particles as they are deflected by a magnet. (Sir J. J. Thomson's apparatus for determining the mass of the electron was a precursor of the device that was brought to its present state of perfection by the 1989 Nobel laureate Wolfgang Paul.) But the number of gold atoms, even in an object as small as a ring, is almost unimaginably large and therefore difficult to measure.

Fortunately there is a strange law of nature that makes the problem tractable. In 1811 the Italian jurist-turned-chemist Amadeo Avogadro, a frail little man with a triangular face and protruding eyes, formulated the law that today bears his name. Avogadro was exceptionally modest, and tended to ascribe even his most creative ideas to others. The father of the chemical atom, John Dalton, had considered a similar hypothesis and rejected it, but this did not prevent the self-effacing Avogadro from giving Dalton the lion's share of credit for its discovery. In any case, the obscurity of the language in which Avogadro described his ideas, combined with his diffidence, kept his profound insight from being recognized by the scientific community for nearly half a century. In science, as in everyday life, it is not enough to be right; you must also speak up.

Avogadro's law states that two identical vessels filled with different gases, at the same temperature and pressure, contain the same number of atoms. This means that in order to set a standard, one need only fill a vessel with gas and count the atoms once. Differences in volume, temperature, and pressure can be accounted for by simple proportions and don't require separate counts. Thus, for example, a second identical vessel, at the same temperature but twice the pressure, contains exactly twice as many atoms as the first. Avogadro's law can even be used to count atoms in solids and liquids, because when such a substance changes its state to a gas, the number of atoms in it does not change.

In this way Avogadro converted the problem of counting atoms into the problem of determining the number of atoms in a single, standard volume, which was selected to be the volume taken up by two grams of the lightest gas, hydrogen. (The gram was chosen because it is a convenient unit, and hydrogen atoms always form pairs—hence two grams.) Avogadro's number is about 6×10^{23}, far above the human capacity to imagine. The immensity of this number is related to the minute size and weight of atoms and accounts for the practical difficulty of using an atom as the standard of mass. If the jeweler who would determine the mass of a ring by counting individual atoms were to count one atom per second, it would take him ten thousand times the age of the universe to finish the chore. In reality this Herculean labor could be accelerated by technological means, but the fact remains that the magnitude of Avogadro's prodigious number stands in the way of adopting the carbon atom as the international standard of mass.

There are other ways to redefine the kilogram. The interlocking of units allows many possibilities, including the suggestion that the new quantum mechanical "practical" units of electricity be promoted to the status of fundamental units and that the kilogram be derived from them. This course recommends itself because of its high precision and is strongly advocated by many scientists. But if it is enacted by international agreement, the concept of mass will lose its connection to the balance scales of primitive farmers, from which its meaning was originally derived. If the second, the meter, the volt, and the ohm were promoted to primary status, a kilogram would become an auxiliary unit. In particular it would equal a second cubed volt squared per meter squared ohm—in other words gibberish to all but the initiated. In terms of a practical device for calibrating mass, this definition would be just as inscrutable: A kilogram is that mass which, when propelled by a motor powered by a one-volt source and having a resistance of one ohm, acquires a speed of one meter per second in precisely half a second.

However it is done, defining the second, the meter, and the kilogram in terms of atoms is about as precise as one can get without losing contact with commonsense notions of measurement. Consider the dimensions of this page. A ruler indicates that it is so many inches long. But under a microscope the edges look as ragged as the coastline of Maine. Where should the ruler be applied? Even the most perfect facet of a crystal, examined by an STM, looks like the surface of a cheese grater, each bump an individual atom. Does the length of the crystal begin at the bumps or in the valleys between them? The question recalls the problem of understanding the meaning of touch in an atomic-force microscope.

The definition of mass on the macroscopic scale suffers from another kind of problem. We must assume that the mass of an object is fairly constant, otherwise the concept of mass ceases to make sense. But every time a finger touches a piece of metal, thousands of atoms change places. If the metal is exposed to the atmosphere, the way Davisson and Germer's nickel target was, its surface gradually collects a film of air molecules and water. (This is the reason why the International Committee on Weights and Measures issued a bulletin in 1989 to explain that comparisons of unknown masses with the primary standard should be made "just after cleaning and washing" both objects, but insisted that this procedure does not constitute a new definition of the kilogram. One does not tamper lightly with international treaties.) Of course, one could keep a chunk of metal sealed in a vacuum and never take it out, but then it could not be measured and would be useless as a standard.

An even more fundamental problem is that, in the subatomic realm, mass is continuously converted into energy, and vice versa, according to the formula $E = mc^2$. This implies that an object, in order to have a fixed mass, must be sheltered not only from loss or accretion of atoms but also from exchanges of energy with its surroundings. A stray beam of light, an undetected source of heat, a mote of radioactive dust from Chernobyl caught in the glass of the container, and even a blast of the horn of a passing car—all these influences can change a fixed mass by a minute amount. At a certain level of precision the concept of mass as a fixed attribute of an object loses its meaning.

Long before that level is reached, however, the kilogram will be redefined with reference to atoms. The metric system, that great instrument of global communication and understanding, will then have passed beyond intuitive comprehension. The second is defined in terms of billions of microwave oscillations. The meter is the distance light travels in a few billionths of a second. The kilogram may be defined as the mass of billions of billions of billions of carbon atoms. (The escalation from a billion to a billion cubed is a reflection of the enormity of Avogadro's number and is ultimately caused by the three-dimensional nature of massive objects, as compared with the one-dimensionality of length and time.) Whereas a million is barely imaginable, a billion is an empty phrase. Worse still, the devices that make measurements—cesium clocks, lasers, and mass spectrometers—are vastly more complicated than pocket watches, meter sticks, and balance scales.

In this light the International Kilogram in Paris doesn't seem so obsolete after all. It is an essential scientific tool and a perfectly ordinary and palpable object at the same time—one of the last remaining links between modern physics and everyday life. There is melancholy in the realization that in the near future this little cylinder of polished metal under its sparkling bell-jars will relinquish its unique function and join the standard meter as a quaint museum piece. The loss of a faithful assistant who has served without fail, albeit passively, for over a century, is sad, even if it is only an inanimate object. The thought of all the human energy and care, not to mention international cooperation, that had gone into its creation and preservation through all these years evokes a sense of impending loss.

By discarding that human-scaled prototype, science will be taking still another step into inscrutability. If, as Einstein remarked, science is nothing more than a refinement of everyday thinking, then loss of contact with common sense is a dangerous trend. If concepts as fundamental as the units of time, distance, and mass are defined in terms of atoms rather than ordinary things, how are we ever to arrive at an understanding of atoms without falling into a trap of abstract, circular reasoning?

Our only hope of progress lies in taming the atom. Atoms are no longer as remote as they seemed a hundred years ago when the

International Kilogram was cast, but at the same time the subnuclear particles of concern to modern physicists are as far beyond the atom as the atom is beyond us. If we are to come to terms with the science of the future, atoms must become even more ordinary, for Protagoras's maxim is about to be revised: In the twenty-first century the atom will replace man as the measure of all things.

11

Large-Scale Quantum Mechanics

The most profound mystery confronting physics at the end of the twentieth century is neatly captured in a Charles Addams cartoon that appeared in *The New Yorker* magazine in 1940. The setting is a wintry landscape. An eerie light casts long shadows upon the pristine snow. In the foreground a crouching skier speeds down a hillside, leaving twin tracks that trail up the slope behind him, diverge to pass on opposite sides of an enormous pine, then rejoin to continue on in normal, parallel fashion. Another skier looks on in amazement.

The power of the cartoon derives from the contrast between what our eyes can plainly see and what our brains know to be impossible. If, instead of a skier, Addams had depicted something altogether different—an avalanche, say, or better yet, a mountain stream—no one would give the scene a second thought. There is nothing strange about a current of water flowing around a tree and reconstituting itself on the other side. But for a solid object to pass through an impenetrable barrier is impossible.

It is impossible in our macroscopic, everyday world, but in the realm of atoms, where quantum mechanics reigns, the rules are different. It is normal for an atomic particle to occupy two places at once, to tunnel through a barrier, or, in variants of Young's double-slit experiment, to circumvent an obstacle on both sides at once. For this reason the Addams cartoon has an immediate appeal to physicists. It tends to crop up on overhead transparencies that are shown to lighten the mood at the beginning of difficult technical lectures about quantum interference and at the end of summary talks on modern developments in atomic physics, as a visual crutch to ease the audience's transition back to the real world. The picture has even been reprinted in a scholarly journal with a paper about the experimental investigation of wave-particle duality. Scientific audiences respond instantly to the uncanny precision with which Addams has unintentionally captured the dilemma of quantum theory: If atoms obey spooky rules, and we are made of atoms, why don't we follow the same rules?

The answer to the quandary must lie on the theoretical ladder that leads from the laboratory down into the world of atoms, precisely at the missing rung between the two regimes, where classical physics loses its relevance and quantum mechanics takes over. Since 1925 generations of physicists have either ignored the gap by simply stepping past it or stumbled over jerry-built attempts to bridge it. The major thrust of most experimental efforts to explore the problem has been to augment the human senses enough to make atoms visible and tangible, in order to arrive at a more perfect understanding of their inner workings. The opposite approach is to amplify quantum effects to the level of ordinary sensory perception so that their strangeness can appear displayed in full view. If this effort succeeds, by the beginning of the twenty-first century we may actually witness a laboratory demonstration of a small version of the miracle captured by Charles Addams.

Most of the atomic phenomena that are specifically quantum mechanical stem from the wavelike nature of particles. Conventional wisdom declares that waves can spread out freely in space, whereas particles cannot and must remain localized in one spot. But in the atomic world, particles can flow over and through each other, pass through forbidden gaps (as demonstrated by scanning tunneling microscopy), and annihilate each other by destructive interference. Young's double-slit experiment, with electrons replacing light, displays this effect more convincingly than any other, which is why Richard Feynman regarded it as the fundamental paradigm, the quivering heart, of quantum mechanics.

When Davisson and Germer discovered the interference of electrons in 1925, they ushered in the study of matter waves, which is to say quantum theory. Interference experiments were soon conducted with atoms in place of electrons, and eventually there was no doubt that all atomic particles do indeed have wavelike characteristics. But experiments like that of Davisson and Germer demonstrate waves only indirectly; they differ from Young's experiment in that the openings through which the particles pass cannot be seen or felt. The slits are interstitial spaces between atoms, not holes in actual screens. Even if atomic particles do behave like waves in microscopic surroundings, the question remains whether they are waves in the usual sense. Do they show interference effects when the slits are macroscopic—ordinary, visible gaps in real walls?

For electrons the answer is yes. Electron waves bend around obstacles and interfere with each other in exactly the same way as light waves. A practical application of this similarity is the electron microscope, which substitutes beams of electrons for beams of light to magnify minuscule specimens. But electron waves are not really matter waves in the sense that electrons are not really particles of matter. Their weight contributes an almost negligible amount to the weight of a body, and as for size, they probably have none at all. They are pointlike particles, with an insignificant bit of mass and a tiny quantum of magnetism, but to call an electron a material object is stretching the point.

Atoms, on the other hand, are rapidly approaching the status of *things*. Even though they are not macroscopic, they are definitely material, with weight and size, and in many senses they can be regarded as objects. The remarkable quantitative successes of quantum theory constitute convincing indirect proof of the waviness of atoms, but until recently it has not been demonstrated directly. The practical difficulty of performing the double-slit experiment with atoms resides in the extreme shortness of their wavelength, which is a consequence of their relatively great mass: A typical atomic wavelength is several thousand times smaller than the wavelength of visible light. Thus the slits and the spacing between them must be much smaller, to a degree achieved only recently by the most advanced nanofabrication techniques.

In the spring of 1991 four different laboratories independently demonstrated the interference of atoms. The first to report was Professor Jürgen Mlynek, who, with his assistant Oliver Carnal, performed a double-slit experiment with helium atoms at the University of Konstanz, on tranquil Lake Constance, between Switzerland and Germany. Mlynek chose helium because it is the second-lightest element, so its quantum mechanical wavelength is large, and because helium, being an inert gas, poses no problems of corrosion or other chemical interaction with the appartus through which it flows. Within weeks of the publication of this work Mlynek and other researchers on both sides of the Atlantic also described similar experiments with heavier, more active atoms, such as sodium and calcium.

The sketch of Mlynek's apparatus might have come from Young's own papers: the experiment itself was a repetition of the original 1803 version, with the crucial difference that the slits were irradiated not by sunlight but by a stream of material particles. A beam of helium atoms encounters a little gold screen in which two slits have been cut and then, about two feet downstream, enters a movable detector that records the arriving atoms one by one. The two slits, instead of being separated by about a millimeter as Young's were, are only a thousandth of a millimeter apart. The picture of the double slit that accompanies Mlynek and Carnal's article was made by means of an electron microscope; electron waves, once a fundamental scientific discovery, now serve as common laboratory tools.

The outcome of the helium-atom experiment, a chart of the number of atoms that arrive at various locations at the far end of the apparatus, shows the typical interference pattern that is the unique signature of waves: a series of regularly spaced maxima, where waves from the two slits reinforce each other, separated by minima, where they cancel out. Whenever physicists see such a pattern—on the wall of Thomas Young's study, in Davisson and Germer's electron-scattering experiment, between the knuckles of their fingers—they know that they are dealing with waves of some sort. The clarity of the interference pattern produced by helium atoms leaves no doubt that they, too, are waves.

The most mysterious feature of the experiment, which astonished even the intrepid atom counter Sam Hurst when he read about it, is the fact that each atom traversed the apparatus alone, uninfluenced by the jostle of other particles. So it is difficult to understand how an atom can avoid certain destinations at the far end, namely those that will, hours later when thousands of particles have been collected, show up as minima. What mysterious influence steers it to one of those spots where its fellows will later congregate at maxima? How do all those distinct, independent helium atoms conspire to produce that characteristic striated pattern? The only known explanation is that each individual atom is a wave that passes through both slits at the same time, and reconstitutes itself on the far side, where it is recorded as a particle.

About a hundred and twenty years elapsed between Thomas Young's demonstration in 1803 that photons are waves and the Davisson-Germer experiment on electrons in 1925. Sixty-six years after that, in 1991, helium atoms marched around both sides of an impenetrable gold barrier. If the accelerating trend in the pace of discovery continues, how close will we come thirty years later, in the early twenty-first century, to seeing a double-slit experiment performed with molecules, or living cells, or even a skier?

The question was asked, not for skiers but for cars, by George Gamow, one of the architects of the Big Bang theory of the origin of the universe, whose charming popularizations of difficult subjects like quantum theory and special relativity have seduced many young people, including myself, into the study of physics. Actually Gamow considered not double-slit diffraction, which means circumventing a barrier, but the related phenomenon of tunneling through a barrier. Both processes are forbidden to ordinary objects, and both depend on the relationship between the wavelength of the object and the size of the barrier. Young's experiment succeeds only when the slits are not much larger than the wavelength that illuminates them, and an STM works on the principle that the gap through which electrons tunnel is not much wider than the electron wavelength.

In Gamow's fable *Mr. Tompkins in Paperback,* the hero, a little clerk in a big city bank, is worried about the tunneling of material objects through brick walls. He wonders about "a car locked safely in a garage leaking out, just like a good old ghost of the middle ages, through the wall of the garage." "How long have I to wait," he asks the professor, until he can witness such a scene?

After making some rapid calculations in his head, the professor replies, "It will take about 1,000,000,000 . . . 000,000 years." The answer is so enormous that even Mr. Tompkins, who is used to large numbers in the bank, loses track of the number of zeros. Today physicists are nowhere near demonstrating tunneling of macroscopic objects through each other, but the observation of interference of atoms has at least brought it into the realm of possibility.

The primary difficulty of witnessing macroscopic quantum effects is the problem of coherence, which is a greater obstacle than an object's large weight and resulting small wavelength. Interference and other wave phenomena are readily seen in water, but only if the number of interfering waves is small. When many independent ripples, each going its own way, interfere with each other, no recognizable patterns can be found. Such is the case with the quantum mechanical waves of macroscopic objects, be they baseballs, cars, or skiers. Each atom has its own independent wave function, and when large numbers of them combine to form macroscopic objects, the result is an incoherent probability map whose wavelike characteristics are impossible to discern.

The only hope of finding quantum effects in large, composite systems is to look for special situations in which the waves happen to behave in an orderly, coherent manner, like soldiers marching in step, and such assemblies actually do exist. The same particles that have been sent through the jaws of Young's experiment individually—photons, electrons, and helium atoms—can also be made to demonstrate quantum effects in immense aggregates. Unlike the interference effects, whose similarity illustrates the essential unity of nature, the collective phenomena differ from each other. The first two have been harnessed in useful devices—lasers in the case of photons and superconductors for electrons—while the third, superfluidity of liquid helium, remains a laboratory curiosity. Like the diffraction of atoms, these are macroscopic quantum effects that lift the strangeness of subatomic behavior up to the perception of the unaided senses.

The least revolutionary of the three phenomena is the cooperation of photons in a laser. While it is true that the production of laser light from individual atoms or molecules is a quantum mechanical process, the final result of those innumerable acts of generation is not startling: it is simply a beam of light of unprecedented purity and coherence, a smooth, unbroken wave, like a musical note with no overtones or admixture of any other notes whatsoever, which could never issue from a real piano. Such a beam of light was not available before the invention of the laser, but it was easily imaginable. Thus, in spite of the ingenuity of its design, the laser doesn't demonstrate a surprising wave phenomenon and therefore cannot help to train our intuitive notions of quantum behavior.

The superfluidity of liquid helium is a much more spectacular and unexpected phenomenon. At extremely low temperatures liquid helium loses its internal friction, which means that a swirl set up in a cup of the fluid will continue to spin forever. Unfortunately the liquefaction of helium is such a difficult and expensive procedure that it will be a long time before superfluidity becomes as widely accessible as lasers and superconductivity. Fortunately, many of the underlying principles at work in this mechanical effect are also illustrated by its electrical analog, superconductivity.

Superconductivity became a household word in 1986, when the temperature at which the effect sets in was abruptly raised from liquid-helium temperatures, which are difficult to reach, to liquid-nitrogen temperatures, which are found in every dentist's office. The phenomenon thereby moved from the research laboratory to the high school classroom, and the popular press began to report that we would soon be traveling on trains elevated by superconducting magnets, watch TV via superconducting satellite dishes, and receive our electrical energy through superconducting cables. Although these predictions may very well come true, the hype obscures a more philosophical and possibly more significant consequence of the superconducting revolution: It will bring large-scale quantum mechanics into the home.

Superconductivity was discovered accidentally in 1911 by the Dutch physicist Kammerling Onnes in the course of his studies of the properties of materials at the temperature of liquid helium. It is the disappearance, at very low temperatures, of all electrical resistance. A current induced in a ring, say the size of a wedding band, that is made of superconducting material such as lead will continue to flow without diminution and without requiring a battery or other source of power. Experimentally this effect has been observed for periods of more than a year, and theory predicts that it will continue for far longer than that. The practical promise of this process stems from the fact that without resistance electrical energy cannot be converted into heat, and therefore no power is wasted. From the fundamental perspective it borders on the miraculous, because all natural processes are accompanied by friction, loss, waste, inefficiency, and dissipation of energy. A current that encounters no resistance, called a supercurrent, sounds suspiciously like perpetual motion. But it is not an alchemist's dream; it exists.

Unlike the delicate experiments that deal with individual atoms, superconductivity is a robust, large-scale phenomenon, as much at home on the factory floor as it is in the laboratory. A superconducting magnet can hold a thousand-pound hunk of iron indefinitely, or at least until its coolant evaporates and resistance sets in again. There is nothing microscopic about superconductivity except its explanation.

The cause of the phenomenon, which was not discovered until forty years after its first empirical observation, is a combination of two factors, one related to the collective behavior of electrons in solids, the other to the rules of quantum mechanics. The swarm of electrons that constitutes an electrical current in metal can be thought of as a multitude of marbles jiggling their way down through a lattice of obstacles in a pinball machine. The obstacles are heavy, positively charged metal atoms, and unlike the fixed islands of a pinball machine, they can be displaced by a fraction of their own diameters from their normal positions. An electron attracts the atoms within its reach and pulls them toward itself. In this way the electron surrounds itself with a diffuse, positive halo. When the temperature is sufficiently low, the random jiggling of the metal atoms is suppressed, and the halos can form especially well. A halo of positive charge can, in turn, attract another electron, with the net result that most of the electrons that participate in the electrical current pair up into couples, without actually touching. It was the unexpectedness of this pairing—after all, electrons are supposed to repel, not attract, each other—that made the explanation of superconductivity so difficult to formulate.

The second part of the theory is a consequence of the quantum mechanical indistinguishability of elementary particles. While electrons in a crowd behave with extreme individualism, each performing its own peculiar motion and occupying its own unique energy level, electrons in pairs do just the opposite. Like photons and alpha particles (which Rutherford identified as the nuclei of helium atoms), electron pairs actually prefer to adopt a common, shared motion, which they all imitate. Photons in a laser display this kind of collective behavior, which manifests itself as a single wave motion of trillions of photons, and so do helium atoms in a superfluid state. Similarly all pairs of electrons in a ring of superconducting wire circle around in the same direction with exactly the same speed and energy and never stray off on their own paths. The result of this perfect cooperation is the total suppression of the disorderly multiple collisions that cause resistance, and the electronic melee becomes an orderly stream around the ring, with electrons lined up like fighter planes flying in formation.

Since they all do exactly the same thing, the billions of billions of pairs of electrons that constitute a supercurrent can be described by a single Schrödinger wave equation and considered as a simple quantum mechanical unit. This shift in perspective—treating a multitude of electrons as a single wave function—renders the phenomenon of superconductivity plausible in quantum mechanical terms. Recall that in nature there are other systems that show no energy loss, like an electron on the bottom of the energy staircase of a hydrogen atom: Niels Bohr imagined that it circles forever without loss of speed or energy, and since it is a charge in motion, it, too, represents an eternal current—just like a supercurrent. Thus supercurrents emerge as precise analogs of electrons in atoms, but they reveal their quantum effects macroscopically. They are large-scale examples of physical systems that defy the laws of Newton and Maxwell and obey those of Schrödinger and Heisenberg.

In a curious way the significance of the relationship between an atom and a supercurrent is reminiscent of one of the great moments in astronomy, the discovery of the moons of Jupiter by Galileo at the dawn of modern science in 1610. Its importance was to serve as a visible example of a revolving system—just like the one Copernicus had proposed on strictly theoretical grounds for the planets. The moons of Jupiter, seen through primitive telescopes, helped persuade the world of the plausibility of the heliocentric hypothesis. Similarly supercurrents, whose effects are visible to the naked eye, furnish a manifest example of the quantum mechanical behavior of imperceptible atoms.

And yet, for all its quantum mechanical nature, a supercurrent is not like Charles Addams's skier. It does not illustrate the interference of two distinct states of motion of a large-scale object like a baseball flying through two different holes in a fence at the same time. No such ghostly behavior has yet been observed in the everyday world.

But there is hope. In 1980 Anthony Leggett, a brilliant and intense English theorist working at the University of Illinois who is renowned for the virtuosity with which he performs long, complex calculations, and who has made important contributions to the modern development of the problem of quantum measurement, described a set of circumstances under which a supercurrent might display macroscopic interference. He imagined a ring of superconducting material, no larger than the tip of a pin. At one point the ring is cut, and a sliver of normally conducting material is inserted in the gap. A supercurrent can flow back and forth in such a ring, a stream of electricity that reverses direction with the frequency of microwaves.

Like all currents, the supercurrent is accompanied by a magnetic field, which points through the ring like a finger through a wedding band, except that it changes direction every time the current reverses. The magnetic field is known as the flux and is subject to the rules of quantum mechanics, which dictate that it can have only certain allowed values called flux quanta—just as the energies in an atom are restricted to a limited number of discrete values. This flux, according to Leggett, is a macroscopic variable, because it can be determined by a measurement of the supercurrent itself, which is certainly macroscopic: it may not be able to support a thousand pounds, but it does rival the strength of electrical currents in ordinary radio circuits.

Leggett designed a ring with dimensions that were cleverly chosen to allow only two possible values for the flux. An external magnetic field is adjusted to supply half a flux quantum through the ring, and since nature abhors such an amount, the system spontaneously responds to rectify the situation. A supercurrent automatically begins to flow in the ring, either clockwise to complete an entire quantum of flux, or counterclockwise to cancel the offending half-quantum. Thus Leggett proposed the construction of a macroscopic version of an atom with only two energy levels and predicted that the flux would jump back and forth between its two allowed values, zero and one, in a quantum mechanical way. This large-scale quantum behavior was as close as he thought he could come to watching a car tunnel back and forth through the wall of a garage. Unfortunately the technology required to manufacture the device had not been developed in 1980 and won't be in place for several years to come.

Six years later Leggett and a colleague returned to the subject in the journal *Physical Review Letters* with an article intriguingly titled "Quantum Mechanics Versus Macroscopic Realism: Is the Flux There When Nobody Looks?" The question alludes to the conundrum of a tree falling in a lonely forest, far from human habitation. In what sense can the tree be said to have made a sound, even if nobody is listening? Realists have no difficulty answering the question, but those for whom human perceptions play a more decisive role in the structure of the world are more troubled. In a similar way Leggett's question goes to the very heart of the meaning of quantum mechanics.

Macroscopic realism demands that the flux in the superconducting ring must have one value *or* the other at any particular time. According to quantum theory, however, it should reflect *both* values at once—it is indeterminate. To be sure, when the flux is measured, only one value is found, but between measurements both states should exist simultaneously. This parallels Professor Mlynek's experiment in Konstanz, in which the helium atom's wave function followed both paths at the same time, even though the atom itself, when caught in the act of passing through the apparatus, can only be in one place at a time.

Since quantum mechanics violates the rules of ordinary experience, it is impossible to find an exact analogy for Leggett's proposal, but even a rough picture is helpful. Imagine a pair of shallow pans, one containing a marble that rolls from side to side, the other holding water that sloshes back and forth. Except for the moment when it occupies the midpoint, the marble is at all times definitely on one side of the pan or the other. The water, on the other hand, is always, to some extent, in both places.

Now consider an experiment in which the weights of the marble and the water are equal, and their respective pans are covered and balanced on fulcrums, teetering like seesaws. The aim of the experiment is to determine which pan contains the water and which the marble, but the researcher's only tool is a crude device that can tell nothing more than whether a pan is tilting to the left or to the right.

Under these conditions a single observation of either pan yields the determinations "left" or "right," but reveals nothing about a pan's content. Similarly one measurement of the flux in

Leggett's ring proves nothing. But if the observer of the pans makes several measurements in quick succession at equal time intervals, he will discover that the motions of the pans are slightly different—that the water, because of its simultaneous presence in both sides of the pan, has an effect on the speed of its pan's oscillation that is different from that of the marble. Without removing the covers and actually looking, the researcher can determine the hidden contents of the pans. According to Leggett, a similar series of timed measurements of the flux in his ring will reveal traces of the wave function's tunneling back and forth between two distinct macroscopic quantum states, which resembles the sloshing of water in a rocking pan.

The arresting subtitle of the 1986 paper ("Is the Flux There When Nobody Looks?") also relates to the conventional interpretation of quantum theory, which holds that physical reality can only be ascribed to observations and measurements, not to abstract mathematical constructs like the wave function. To make an assertion about what happens in a quantum system between measurements is idle conjecture because, in order to check the assertion, a measurement has to be made. Indeed, we must assume that the quantum object—the flux through a ring, for example—does not exist between measurements, even though, to the extent that the flux, like a chair, is an ordinary macroscopic thing, common sense tells us that it exists even when it isn't being observed. Leggett's question is the fundamental question of objective reality, the "to be or not to be" of physics.

Leggett's interpretation of the proposed experiment was soon challenged by a number of physicists. The chief objection centered on the problem of measuring the flux, which he assumed could be measured without disturbing its value. While this can usually be done for macroscopic quantities, Heisenberg's uncertainty principle may intervene in the case of quantum systems. In certain cases taking a measurement changes the value being measured and renders any prediction invalid. The specific way in which this happens depends on the particular experimental arrangement, and so the controversy about Leggett's proposal could not be resolved until someone described an actual measuring device for the flux and analyzed the details of its operation.

The challenge was taken up three years later by Claudia Denke Tesche and her collaborators at the laboratories of the IBM Research Division in Yorktown Heights, New York. Claudia Tesche is an outgoing and thoroughly practical young woman. As a graduate student at Berkeley, she started working within a branch of particle physics called axiomatic field theory, a subject so abstract that it is sometimes derided as recreational mathematics. However, she soon gave it up in favor of a field more suited to her outlook—to devices that harness the power of quantum mechanics for useful purposes, and in this pursuit she ended up at IBM.

When Tesche learned of Leggett's experiment, its technical difficulty intrigued her as much as its fundamental significance. In 1990, ten years after his initial proposal, she made a conceptual breakthrough by inventing a complicated electronic circuit consisting of miniature superconducting switches and magnetic-field detectors that circumvents the problems posed by the uncertainty principle. Her theoretical calculations predict that with this novel arrangement she will be able to measure the flux in a superconducting ring without destroying the possibility of further measurements later on. Leggett's experiment now seems feasible, and Tesche is engaged in preliminary preparations for its execution.

She believes that the actual apparatus will be completed sometime before the turn of the century, but she's too experienced in the vicissitudes of scientific research to announce a precise date. It will certainly bear no resemblance to Charles Addams's hibernal scene, but will probably consist of an ungainly vacuum flask bathed in liquid helium, surrounded by its thermal cocoon and concealing some nearly invisible electronic circuits strung together in a spiderweb of wires. A tangle of cables and tubes will trail from the vessel to a roomful of humming refrigerators, vacuum pumps, recording devices, and IBM computers. The experimental results will be encoded in strings of seemingly meaningless numbers that refer to measurements of invisible fluxes at preselected time intervals. The whole setup will look decidedly humdrum.

But if the experiment succeeds, its result will be dramatic, no matter which way it comes out. If the sequence of measurements of the value of the flux can only be matched by a quantum mechanical calculation that assumes that the flux through the ring always points in *both* directions at once, a macroscopic system will have been found in an indeterminate state for the first time. If, on the other hand, the measurements indicate that the flux behaves classically, which is to say that it points in one direction *or* the other, quantum mechanics will have failed, and the reason for the failure will have to be found. (One possible source of such a failure is the assumption that the supercurrent is described by a simple Schrödinger equation: even a minuscule influx of heat into the superconducting ring from its surroundings might modify this equation in an uncontrollable way and render its predictions invalid.)

It is no accident that this latest advance in quantum research is not being pursued at a university, but, like the Davisson-Germer experiment at the Bell Telephone Company, in an industrial laboratory. Industry is becoming increasingly interested in this subject. The Hitachi Company's central research laboratory in Tokyo is the site of a continuing series of conferences on the relationship between advances in microelectronics and the foundations of quantum theory. The scientists who attend it come from such firms as IBM, AT&T, and NT&T—Japan's major telecommunications company—to discuss everything from atomic diffraction to superfluidity and superconductivity. What is remarkable about these meetings is not the idea that technical innovation might shed light on fundamental questions—technology has been a key to scientific progress since antiquity—but that they could be held in such a prosaic venue. Holding a symposium on the meaning of quantum theory at an industrial laboratory seems a little like conducting a colloquium on the nature of playfulness as a half-time feature at the Super Bowl. Questions about the interpretation of wave functions would appear to be of no interest to corporate giants concerned with production schedules and profit margins.

But the fact is, what was once a philosophical debate among the high priests of physics is becoming a bread-and-butter issue. Ordinary scientists in the industrial laboratories of the world are starting to wrestle with the meaning of quantum mechanics. Until they understand whether the flux actually exists in a particular superconducting device, they cannot fully explore its practical applications. Technological development, which has profited richly from the well-oiled machinery of quantum theory, has come up against a barrier, making it necessary to reach inside the machine to see what makes it tick.

A fresh burst of energy, fueled by industrial competitiveness and product innovation, is propelling the quest to comprehend nature to new heights. What physicists, spoiled by a half century of the unqualified successes of quantum mechanics, have tried to avoid—addressing the mystery of the meaning of their enterprise—has been forced upon them by technological progress. This pressure may eventually lead to the result that skiers tunneling through pine trees will be regarded, if not as commonplace occurrences, at least as mere extensions of the intuitively acceptable behavior of ordinary matter.

Claudia Tesche is too practical to indulge in speculations concerning the nature and meaning of reality. "*Meaning* is a philosophical word," she shrugs, and then turns her attention to a leaky vacuum pump. But the search for the meaning of quantum mechanics continues to gather momentum, and she is right in the center of the action.

12

In Search of the Missing Rung

On a fine summer's day in August 1834 a heavily laden barge was gliding down the narrow Union Canal near the city of Edinburgh, Scotland, under tow by a team of horses driven by a scruffy postilion. Not far behind rode a young gentleman whose sharp, finely chiseled features gave him an air of hawklike concentration, which seemed to be fixed on the turbulent swirl of the waves in the boat's wake. Suddenly the barge stopped—whether it had hit an obstacle or something had broken was not immediately evident, but the horseman, a twenty-six-year-old scientist and future naval engineer named John Scott Russell, paid no attention to the mishap. Without taking his eyes from the water, he galloped ahead to the front of the vessel and peered down at the waves churning up at its prow.

Later he described the astonishing sight he found there. When the barge stopped, the waves it had created did not stop with it, but kept moving, "assuming the form of a solitary elevation, a rounded, smooth and well defined heap of water, which continued its course along the channel apparently without change of form or diminution of speed." He followed the wave, which was thirty feet long and a foot and a half high, at a speed of "eight or nine" miles an hour for a couple of miles before he eventually lost it in the windings of the channel.

"The Great Wave" and the "Wave Par Excellence," Russell called that impressive phenomenon in his subsequent writings.

Today, under the more prosaic name of solitary wave, it is the subject of continuing theoretical and experimental research by physicists, mathematicians, engineers, and computer scientists, and has found applications wherever waves show up. For example, transatlantic telephone cables may soon be replaced by optical fibers carrying solitary waves of visible light across thousands of miles "without change of form or diminution of speed," and the oscillations of supercurrents in certain ring-shaped electronic devices also turn out to be examples of Russell's wave.

Solitary waves differ profoundly from conventional waves. When a sudden commotion is created on the surface of a lake by, say, a falling log, an ordinary wave will travel away from the source and then quickly flatten, spread, and disperse. Within a short time no trace of the disturbance remains. A solitary wave, on the other hand, persists for long periods of time and travels great distances without changing shape. Its form and speed depend on the details of its surroundings, such as the width and depth of the Union Canal, and its appearance is highly unusual, as the perceptive Russell realized when he encountered it. His chance observation on that August day led him to the discovery of a new type of wave, which, in most unfluidlike fashion, preserves its shape and identity like a solid body.

Such a phenomenon raises interesting possibilities for quantum mechanics. When Louis de Broglie suggested that electrons are associated with waves, he did not know what the waves were made of, nor what shape they assumed. Schrödinger and Born soon showed that they could be interpreted as waves of probability and that their appearance, if it could be followed by eye, would be entirely different from the appearance of a particle. An electron speeding freely through space like a bullet, for example, is associated with a wave function that extends throughout all space like an endless train of rollers spread over the entire ocean, while an electron in an atom—which had been pictured as a miniature planet in Bohr's obsolete theory—is described by a wave that resembles the concentric circular ripples on the surface of a glass of water. The conceptual chasm between the images of an electron as a particle and as a wave would be considerably narrowed if that wave had a more particlelike appearance—if it had a definite position, a well-defined speed, and a compact shape, like Russell's great wave.

The idea that nondispersing waves offer an intuitively appealing compromise between the two conflicting aspects of electrons goes back to the very beginning of quantum theory and continues to this day as a persistent theme in theoretical speculations about the true nature of particles. Recently such phenomena have even begun to show up in experimental studies of the atom.

When, in a breathtaking burst of scientific creativity that irrevocably altered the way we perceive the world, Erwin Schrödinger developed his wave equation in the spring of 1926, he was searching for a more physical picture of the atom than Heisenberg had provided. Heisenberg's guiding philosophy, which did not allow him to make use of unobservable quantities, had led him to a powerful but exceedingly abstract formulation of quantum mechanics. In the second of his six great papers on wave mechanics, Schrödinger took aim at Heisenberg's approach:

> It has even been doubted whether what goes on in an atom can be described within the scheme of space and time. From a philosophical standpoint, I should consider a conclusive decision in this sense as equivalent to complete surrender. For we cannot really avoid our thinking in terms of space and time, and what we cannot comprehend within it, we cannot comprehend at all. There *are* such things, but I do not believe that atomic structure is one of them.

He did not explain what he meant by "such things" that cannot be described in spatial and temporal terms, but perhaps he had in mind abstract physical concepts such as electric charge and the magnetic field; in any case he left no doubt that he believed atoms should be described in a more picturesque way than by matrices.

In a letter to a colleague he made this point even more explicitly:

> That a space-time description is impossible, I reject totally. Physics does not consist only of atomic research, science does not consist only of physics, and life does not consist only of science. The aim of atomic research is to fit our empirical knowledge concerning it into our other thinking. All of this other thinking, so far as it concerns the outer world, is active in space and time. If it cannot be fitted into space and time, then it fails in its whole aim and one does not know what purpose it really serves.

Atoms, in other words, should be described in terms that relate them to our perception of the everyday world.

Unfortunately the waves with which Schrödinger succeeded in describing the interior of the atom, and which he himself proved to be logically equivalent to Heisenberg's laundry lists of numbers, do not resemble our experience of electrons in the least. They do not show even a trace of the pointlike particle nature of electrons that J. J. Thomson had discovered and that is revealed by electronic and photographic recording devices. In view of the fact that he had initially rejected the suggestion that electrons are waves, it is ironic that Schrödinger himself was responsible for this paradox.

The problem of reconciling the two sides of an electron's nature is one facet of the more general problem of reconciling quantum mechanics with classical mechanics. If two theories both apply to the same object, they should have clearly defined domains of applicability, and the boundary between them should be comprehensible. Einstein's special relativity, for example, applies to objects such as baseballs and rocket ships, and so does classical Newtonian mechanics. The connection between the two theories is straightforward and is provided by a single quantity: the particle's speed. When a body moves slowly, Newton is in charge, but as it accelerates to a velocity near that of light, Einstein takes over. The formulas of the two theories join together seamlessly in such a way that as the speed variable changes, they change automatically from one mathematical form to the other, while at intermediate speeds, where both theories apply to some degree of approximation, the formulas agree with each other.

The transition from classical mechanics to quantum mechanics is much more murky. In spite of the best efforts of physicists since 1925, no general relationship like the simple connection between mechanics and relativity has yet been discovered. The quantum-classical boundary is a no-man's-land that physicists navigate more by intuition born of experience than by reason, using quantum mechanical results where they fit, and classical mechanics where it seems more appropriate. There is only a small handful of special circumstances in which the demarcation between the two descriptions of nature can be studied in an orderly manner.

A hydrogen atom excited to a very high level of energy provides one such example. From the point of view of Bohr's old theory, the electron would move slowly around the nucleus at a large distance, and there, like a distant planet, follow a well-defined elliptical trajectory. In this orbit its motion would be described according to classical Newtonian mechanics like that of the electrons in J. J. Thomson's tube and Hans Dehmelt's trap. However, in quantum mechanical terms the atom looks very different. The wave function for the electron in this high energy level is very well known—a college junior can compute and plot it. It represents a probability map that is spread out over the entire atom, and a graph of its magnitude resembles the pattern of ripples caused by a raindrop falling on a pond, except that the highest crests would be near the outer rim of the atom, rather than in the center. Schrödinger hoped that the reconciliation of this view with Bohr's particulate conception would help to clear up the relationship between quantum and classical mechanics.

To achieve such an accommodation, he searched for a compromise, a phenomenon that is fundamentally wavelike but also displays some of the features of a particle. Russell's solitary wave would have served splendidly, but unfortunately it was not applicable. The mathematical form of a solitary wave happens to be incompatible with Schrödinger's equation, but there is another type of wave form, called a wave packet, which looks deceptively like a solitary wave, and in addition fits smoothly into the theory. He thought that at the edge of the atom a number of quantum waves might interfere with each other and conspire to form a single tight wave packet, which would travel around the nucleus in a Newtonian elliptical orbit. Such a packet constructed from the waves described by his equation would represent a marriage of the old mechanics with the new: Waviness would reside in the fundamental description of the atom, and particulateness in the isolated wave packets, which revolve around the nucleus like planets.

As soon as Schrödinger had finished the formidable task of developing his version of quantum theory in June 1926, he wrote a letter about his wave-packet conjecture to the Dutch professor Hendrik Lorentz, who had shared the 1902 Nobel Prize in physics for his explanation of how magnets affect atoms and at seventy-three years of age was regarded as the dean of theoretical physicists. Schrödinger emphasized the urgent need for a theoretical construction of "wave groups (or wave packets), which . . . mediate the transition to macroscopic mechanics." But he despaired of ever pinning them down because of the "great computational difficulties" he was encountering. Wouldn't it be nice, he added wistfully, if a calculation could be carried out not just for the hydrogen atom but for all quantum waves in general? But he knew that this was a hopeless wish, at least for the moment.

The problem, as Lorentz replied immediately, is that the mathematical form of the Schrödinger equation dictates that all conceivable atomic wave packets inexorably spread out with time. They may move along a definite trajectory temporarily, but they soon disperse and lose their shape and cohesiveness. And once dispersed, Lorentz continued, you could hardly expect them to reassemble again into tight bundles. Wave packets behave, in other words, like normal waves, not like Russell's great wave, or like particles. Lorentz realized that a nondispersing wave, for all its simple appearance, is a subtle phenomenon that depends on the delicate interplay of influences that are not described by a standard wave equation like Schrödinger's. In the interval between letters Schrödinger himself had come to the same melancholy conclusion on his own.

And there the matter rested until Schrödinger's death in

1961, and for another generation after that. Most physicists regarded the wave function as a magical recipe for predicting probabilities that happened to facilitate the design of better electronic devices and faster lasers. Graduate students were taught Born's probability interpretation of the wave function—that the wave function determines the likelihood of an electron being found at a given spot—as revealed gospel, and were told that the transition to classical physics, while not fully understood, is a technical detail without practical consequences. Such phrases as wave-particle duality were declared obsolete relics of bygone days, and although a few die-hards kept alive Schrödinger's dream of finding a picturesque description of the atomic interior to replace Heisenberg's dry catalog of numbers, little progress was made. Success lulled physicists into pragmatic acceptance of the received doctrine.

Now, in the final decade of the twentieth century, more than sixty-five years after Schrödinger's miraculous spring of 1926, two new techniques are reviving interest in the subject of wave packets and their role in atoms. Theory is benefiting from the power of high-speed computers, and fast-pulsed lasers are beginning to make a new class of experiments possible. Again, just as in the case of organic dye lasers that record the birth of organic molecules and superconducting rings that monitor macroscopic quantum effects, the technological offspring of quantum theory are pressed into service to help uncover the meaning of their own roots.

The modern computers that have begun to generate pictures of wave packets within hydrogen atoms would have been a delight to Schrödinger. In 1988 a team headed by Carlos R. Stroud, Jr., at the University of Rochester reported that it had used Schrödinger's own solutions of the wave equation and combined them to produce a series of computer images. The difference between this accomplishment and Schrödinger's own abortive efforts is simple: computational power. Adding hundreds of complicated terms, and repeating the process over and over to follow the packet's hypothetical motion around the nucleus, is child's play for a computer; displaying the results in graphical form for visual impact is routine. Even a computer-generated movie of a wave packet in action would not be difficult to make and would be a helpful guide for our imagination.

According to the new computer images, Schrödinger's hope was justified: The wave packet does follow a planetary trajectory. Like a heap of water, the tightly packed lump of probability begins at some point far away from the nucleus and, without losing its shape, pursues an elliptical path traveling at the same speed that an ordinary particle would have under the same circumstances. Studying that wave's orbit around the atom restores one's faith in the unity of physics: Newton's celestial mechanics, Bohr's old-fashioned but graphic theory of the hydrogen atom, and Schrödinger's revolutionary quantum mechanics come together and agree with each other in a most satisfying way.

In the end, choosing between the two images—the cloudy blur of waves or the miniature solar system—simply depends on how the system is prepared. Treat an electron like a particle, and it will produce a dot on a photographic plate. Manipulate it as though it were a wave, and it will leave the telltale marks of interference. The appearance of the inside of an atom is determined by how you look at it.

Until recently the tools available for reaching into the atom, such as Wineland's ultraviolet light beams that induce quantum jumps and Rutherford's alpha particles that penetrate all the way to the nucleus, have been blunt instruments that either revealed little detail or mangled the delicate structure of the atomic interior beyond recognition. The issue wasn't so much a matter of overbearing force—the intensity of a lamp can be turned down to feeble levels—as of insufficient speed. Electrons whirl around inside the atom at speeds approaching the speed of light, so any device designed for imaging them has to be quick about it. The watchmaker who holds his screwdriver in the gears is sure to ruin the mechanism, but if he delicately touches a wheel here and a spring there for a fraction of a second, he may uncover the secrets of the watch without destroying it. Picosecond laser pulses are proving to be the deft pointers that allow us to probe the interior of the atom.

They were the experimental tools Carlos Stroud used in order to explore the same wave packets in real atoms that he had simulated in a computer. First he and his team raised an atom of sodium into a very high state of excitation by means of a finely tuned burst of microwaves. The outermost electron was thus carried into an orbit that is thousands of times larger than it would normally be. At that distance the electrical force of attraction to the nucleus is very weak, so the electron moves much more slowly than usual—just as the outermost planet, Pluto, moves only one sixth as fast in its orbit as Earth.

In the second step of the experiment Stroud's team hit the atom with a fast laser pulse. The pulse fixed the location of the electron at a particular spot in its distant orbit and reorganized its enormous, complicated wave function into a small wave packet centered on that spot. The process resembled an act of measurement: When an electron is detected, say at a particular point on a TV screen, the wave function, and thereby the probability of finding the electron, collapses to that point. The wave packet that Stroud actually created by use of the laser corresponded to one that his computer had previously generated by adding a large number of ordinary atomic wave functions.

After the atom had been prepared in this way, it was left to its own devices for a fraction of a microsecond. During this interval it can be described in two equivalent ways: In the language of classical mechanics, the electron revolves at a snail's pace on a huge elliptical path around the nucleus in a course of motion that could have been determined by Isaac Newton. The second description requires a digital computer to follow each of the constituent quantum waves, all of which spread over the entire miniature solar system of the sodium atom. At the beginning they all add up to a single small wave packet at the location of the electron. Then each one evolves on its own according to the Schrödinger equation. Step by tiny step the computer follows the twists and wiggles of each undulation in a sequence of intricate mathematical operations that would be impossible to duplicate by hand. Finally, when all the waves have been put together again, most of the troughs and crests miraculously cancel each other out, leaving only a little packet where it should be at the advanced position of the electron in its orbit.

In the final step of the experiment Stroud tore the electron out of its atom by means of an external electric field. Since it is held in place by the force of its attraction to the nucleus, it can be removed by an opposing electrical force applied by letting the atom pass near a positively charged metal plate. The ease with which this could be accomplished revealed the approximate position of the electron in its orbit at the moment of ejection: At perigee, when it is closest to the nucleus, it is much more tightly attached than at apogee, when it is farthest away. Lacking a microscope with which he could follow the wave packet directly, Stroud resorted to this trick to determine its final position indirectly.

The agreement of the data with the predictions of both the Newtonian and the quantum mechanical theories vindicated Schrödinger's hunch and at the same time revived Bohr's picture of the atom, sixty-five years after it had been declared obsolete. In Stroud's experiment the classical picture of the electron speeding around the nucleus is mathematically equivalent to the quantum mechanical fuzzy cloud, but in practice far more efficient. As Schrödinger had predicted, a single wave packet moving like a planet can indeed replace the combined development of a large number of quantum waves. Schrödinger's conjectured compromise between classical and quantum physics has at long last received experimental confirmation.

But this belated corroboration of Schrödinger's physical intuition does not resolve the wave-particle paradox. Bohr's model has been reinstated only as a helpful calculational shortcut in one special circumstance, not as a general theory of the atom. At the very outset of his calculations Schrödinger had convinced himself, and Lorentz had agreed, that wave packets in atoms cannot cohere indefinitely but must inevitably disperse—and they were right. In Stroud's laser experiment the wave packet followed its elliptical path for only a fraction of a revolution, but in his computer simulations the same wave packet was followed through many successive trips around the nucleus. And indeed, after a few revolutions it begins to disintegrate. A dozen revolutions later it has lost its original shape, and the probability of finding the electron is spread evenly around the entire orbit. The atom has reverted spontaneously to its wavelike aspect, and it is almost as if we have seen the transformation of a particle into a wave before our very eyes.

For a realistic space-time description of the atom this is bad news, for it emphasizes once more that a wave, or even a tight packet of waves, cannot really *be* the electron, as Schrödinger had hoped. But the disintegration of the wave packet is not the end of the story. The computer that followed the process continued grinding on hour after hour in calculations that for sheer volume far surpass human capability. The sequence of pictures published by Stroud's team looks like photographs of water waves in a ring-shaped channel. It begins with an image of a tall, narrow wave packet launched on its course like Russell's solitary wave, but after a dozen revolutions nothing much is left of it but a few evenly dispersed ripples. Then, by the time the particle would have made many more revolutions, the motion of the wave becomes violent. Mysterious peaks suddenly crop up and travel around at the original speed of the packet, only to subside again just as quickly. At a time corresponding to just over a hundred circuits, the wave begins to gather its strength and to regroup in one place. And then the original wave packet, only slightly broadened, appears again at the right place and speed, as if nothing had happened in between. In the words of the authors, the wave packet has decayed and then revived, in direct contradiction to Lorentz's intuitive expectation.

Although the original wave packet does not recover perfectly, and each subsequent revival is less sharply delineated than its predecessor until eventually there is no trace of a wave packet left, the regular sequence of decays and revivals is an exciting discovery. That the electron, like the platypus, has a dual nature has been evident since the birth of quantum theory, but heretofore we have seen only one or the other of its two faces at a time. Stroud's electron, on the other hand, switches back and forth between its two personalities, revealing one, then the other, before finally settling down. It is a signal that we have reached the threshold between particle and wave.

In the fall of 1989 Professor Stroud's team actually crossed it. In a paper entitled "Observation of the Collapse and Revival of an . . . Electronic Wave Packet," they reported on a refinement of their first experiment in which they managed to follow an electron through a large number of revolutions and found the unmistakable imprint of the electron's vacillating nature. The original wave packet disappears, so that the physicist trying to follow the atom's evolution must resort to a quantum mechanical description, but then it regroups in such a way that classical mechanics can take over again.

This work goes beyond an illumination of the missing rung between quantum theory and classical mechanics; it represents a progression from observation to experimentation, from anatomy to surgery. When ultraviolet light illuminates a mercury atom in David Wineland's trap, it promotes internal electrons to higher levels of energy. Their wave functions acquire different forms, which are dictated by nature, not by the experimenter, so the experiment amounts to a passive observation. In Stroud's experiments, on the other hand, pulsed light rearranges the appearance of the wave function in a predetermined way. It is as though we have learned to mold wave functions like clay and watch their slithering gyrations. In this forcing of natural phenomena lies the difference between modern physics and medieval philosophy. When Sir Francis Bacon extolled the power of the nascent scientific method in the beginning of the seventeenth century, he defined its aim as "putting nature to the question," by which he meant torturing her into giving up her secrets. The formation of wave packets by means of laser pulses represents the extension of this method to the hidden world inside the atom and promises a rich harvest of fresh insights yet to come.

Stroud's experiments have opened up a new phase in atomic research. Picosecond pulses will inevitably be followed by femtosecond pulses, and supercomputers by hypercomputers. The manipulation of atoms will make way for the manipulation of atomic constituents—and not in isolation, as heretofore, but in vivo, as it were. Eventually the interior of the atom will become tamed in the same way that the atom as a whole is becoming familiar by means of Pauli traps and scanning tunneling microscopes.

No matter what questions are put to the atom, it is likely that the Schrödinger equation, solved with the help of superhuman computers, will be able to answer them, so that John Scott Russell's dispersionless wave will probably play no role in such explorations. But it is too impressive a phenomenon to be wasted on barge canals and transatlantic telephone cables; to the theoretical physicist it represents a beautiful solution in search of a problem, and speculations about its role in fundamental physics continue to crop up.

Most recently the motivation for these efforts has been not so much the desire to understand the atom itself but Einstein's idea that a particle and its surrounding force field—the earth and its gravity, an electron and the electrical field it causes, a nuclear constituent and the strong force it exerts—should each be described as a single entity. This suggestion has not yet been implemented, but if it is, a promising candidate for its underlying mechanism is a solitary wave (in view of the fact that the Schrödinger-Lorentz objection does not apply to the complicated equations which govern fields). In such schemes the field can be imagined as a vast, smooth ocean and the particle as a solitary wave, a welling-up that retains its shape and speed as it travels across the surface. When it encounters another particle, the two interact briefly, as water waves do, and then both speed off again. In this beguiling image the dichotomy between particles and fields has been eliminated—another giant step in the program of unification of the foundations of physics that began with the atomic doctrine of Leucippus and Democritus. As Einstein noted, a space devoid of particles would contain no fields, in this theory, so the vacuum, too, would be simplified. Except for its mathematical complexity, this model of the world is a tantalizing prospect for theoreticians.

Unfortunately the image of the electron as a bulge in the electric field, which extends outward from it like a vast, tenuous halo, meets a devastating objection right at the outset. Whereas Russell's wave had a definite shape and size, the electron is believed to be a dimensionless point; no experimental evidence for a finite electron radius has ever been found. On the other hand, a pointlike solitary wave is both mathematically impossible and intuitively absurd. Of course it might turn out that the size of the electron is indeed finite, but much too small to be detected with today's instrumentation; if that is true, as some physicists hope, then a solitary-wave theory of the electron will become a viable proposition. In the meantime, however, there are plenty of other particles that lend themselves more naturally to such speculations.

A seminal contribution toward the realization of Einstein's vision was made by Ed Witten, of Princeton University, who is not well known in public but who is regarded by his colleagues as arguably the world's most brilliant theoretical physicist on account of his prodigious creativity and unmatched mathematical wizardry. In 1983 Witten proposed that a solitary wave might solve the stubborn problem of the nature of the proton. The proton—the nucleus of a hydrogen atom—has a diameter of about a fermi (femtometer) and consists of quarks, which hold each other together by means of a nuclear force that is much stronger than gravity, electricity, or magnetism. What is unclear is why a proton cannot be dissected like an atom so that its constituents become isolated like the nuclei and electrons that compose atoms. As a possible explanation, Witten sketched the outlines of a mathematical model in which the proton is a solitary wave in an unobservable underlying sea of quarks. It resists disintegration by a subtle interplay of forces reminiscent of the mechanism that ensured the integrity of Russell's wave for two miles down the Union Canal. For now, and in the near future, Witten's idea must remain a conjecture because quantum chromodynamics, the currently accepted theory of quarks and their mutual interactions, is not yet sufficiently tractable to test its validity. Perhaps proton solitary waves will share the fate of electron wave packets, which were not constructed until more than half a century after Schrödinger envisioned them but then displayed properties far stranger than any he imagined.

In his first letter to Schrödinger, dated May 27, 1926, Hendrik Lorentz wrote that although wave packets are not the ultimate solution to the riddle of matter, he nevertheless hoped that they would contribute significantly to the effort of "penetrating more deeply into these mysteries." Recent progress has justified this hope, for wave packets, and especially solitary waves, do indeed succeed as satisfactory conceptual stepping-stones from waves to particles. But even if the electron, the proton, and the atom as a whole were described by mathematical functions that resemble mounds of water traveling down a canal without spreading out, the fundamental mystery of the atom would remain: Are those solitary waves real, or does nature merely behave as if they are?

13

Quantum Reality

As a graduation gift from college my father gave me two historic photographs, which he had inherited from his uncle, Otto von Baeyer, a prominent experimental physicist whom I barely remember because he died when I was eight. They are now hanging over my desk, mounted in matching wooden frames, one above the other. Both are sepia prints from the glory days of physics in my native city of Berlin—the early 1920s. The upper one is arranged around Albert Einstein, who sits sideways on the arm of a settee at the far left of the picture, facing an imposing assemblage of colleagues, including Otto Hahn and Lise Meitner, who would later gain fame for their roles in the discovery of nuclear fission. With his wild mane of hair still black, his bushy mustache, and those round eyes that manage to appear mournful and at the same time amused, Einstein dominates the room. The lower picture shows a number of younger people facing the camera as they crowd around a shy Niels Bohr with his hands behind his back and a pixieish smile that makes him look more like a young graduate student than a renowned theorist.

My granduncle happens to occupy a prominent spot in the middle of the back row in both pictures, probably because of his exceptional height, but the obvious centers of attention are Einstein and Bohr, who earned the 1921 and 1922 Nobel Prizes respectively. (Appropriately, they were awarded jointly in 1922.) Since relativity and quantum mechanics provide the underpinnings for much of modern physics, the two men are regarded by theoretical physicists as personifications of the intellectual foundations of their profession.

At the time the photographs were taken, the quantum theory of Heisenberg, Schrödinger, and Born was just over the horizon. It was to drive a deep philosophical wedge between Einstein and Bohr. Although the two men continued to hold each other in the highest regard, there would come a time, two decades later, when Einstein would write about his relationship to Bohr's disciples: "We have become antipodean in our scientific expectations," referring to the vast distance that separates Bohr's probabilistic interpretation of quantum mechanics from his own deterministic world picture. Today the subterranean fault line between the two positions is beginning, after decades of quiescence, to rumble ominously, although the majority of physicists ignore or deny it.

In hindsight one can detect hints in my two photos of what was to come. The group of scientists posed with Einstein presents a picture of German academic solidity. It is taken in front of a window with drawn curtains and an enormous leather sofa in which the tiny Lise Meitner almost sinks out of view. It suggests settledness, a sense of reliance on established authority and tradition—an image of orthodoxy. The other group, in contrast, is shown on an unkempt lawn outdoors in front of a building and everyone is standing up. A window with drawn blinds in the background might be the same one that is shown in the upper photograph, seen from the outside. The young scientists around Bohr look ready to follow him into an enticing future. A caption on the back of the photo identifies the group as the "Colloquium Without Bigwigs," or what might be called the Young Turks, who had organized an opportunity to talk to their famous guest without interference from the senior professors. To me the picture represents the coming revolution.

The issue over which Einstein and Bohr clashed was the philosophical problem of realism. Both men understood and accepted the practical successes of quantum theory, and neither had any difficulty with the mathematics, but they differed in its interpretation. The gist of their disagreement can be stated simply. Bohr maintained that since wave functions are usually spread out over the entire atom, electrons in atoms do not have sharply delineated trajectories in space and time, the way planets do. Einstein felt that they must, although he was not able to square his intuition with the principles of quantum theory.

Young's double-slit experiment performed with electrons, which is Feynman's paradigmatic example of quantum behavior, illustrates the dilemma. The facts are well known: When electrons, or any particles for that matter, are fired one by one at two slits, they will pass through and slowly build up a striated interference pattern on a distant screen. Quantum mechanics describes the pattern in precise mathematical detail. When one of the slits is blocked, interference stops, and the pattern on the screen becomes a perfectly ordinary image of the open slit. Furthermore each electron that arrives at the screen can be detected as a particle—causing a minute scintillation, a click on an electronic counter or a silver dot on a photograph.

In the standard description of the phenomenon, called the Copenhagen interpretation, in honor of the city where Bohr established his school of theoretical, phisics each electron is associated with a wave that flows smoothly through the two slits and bunches up into maxima and minima in the manner of all waves. This much is clear and unambiguous. But at the last stage a miracle occurs. When the particle is detected at the screen, the wave function abruptly ceases to exist—or "collapses," in the jargon of the trade. The electron switches from a potential location to an actual one corresponding to any one of a large number of allowable sites determined by the wave function, and then manifests itself as an ordinary minuscule particle. Following Max Born's prescription, the probability of finding the electron at a given spot on the screen is accurately measured by the numerical value of the wave function at that spot.

The wave function itself obeys the Schrödinger equation, while the operation of an electron counter, or any other mechanism for measuring the position of the electron, proceeds according to the classical laws of physics. But the instant in which the wave disappears and the particle materializes, the moment of wave-function collapse, the point at which quantum mechanics makes way for classical mechanics—that wrenching leap from potentiality into actuality—is not covered by any theory. It is euphemistically called the measurement act, and, unlike the steps that precede and follow, has no mathematical description. Although it has led to volumes of abstruse theory and cascades of convoluted debate, this critical moment remains an enigma.

The measurement problem represents an unfinished piece of business for quantum theory, but it was not the principal source of disagreement between Einstein and Bohr. Their dispute concerned the meaning of the wave function as it developed evenly from its source, through the slits, to the screen. Bohr maintained that since a single particle cannot pass through both slits simultaneously, the notion of a particle trajectory between actual sightings makes no sense. The electron is potentially either a particle or a wave, and the apparatus brings out its character: When it is passing through the slits, it is a wave, when it is caught, it is a particle. An electron in an atom behaves in the same way: It is neither a particle nor a wave; it is endowed with neither position nor velocity, until one of the possibilities is fixed by means of a specific macroscopic apparatus. An atom, according to Bohr, represents a different reality from that of the ordinary world of our sense perceptions, and it is unreasonable to insist on forcing the language of our familiar macroscopic surroundings onto that alien mode of existence.

Einstein rebelled against the fundamental obscurity of the Copenhagen interpretation. He felt instinctively that underlying the uncertainty of quantum mechanics there is an objective reality, a more complete description of events in space and time, that has not yet been discovered. From the beginning of the quantum theory until his death he engaged in a futile campaign to convert Bohr to his point of view. In conversations at scientific meetings, in private letters, in occasional publications, such as the paper he helped to write in 1935 under the title "Can the Quantum-Mechanical Description of Physical Reality Be Considered Complete?" (Einstein thought no) and Bohr's published reply (yes), the two giants wrestled for decades.

With respect to Young's experiment, Einstein wanted to believe that the electron really passes through one slit or the other. Bohr countered that as soon as you check, you destroy the wave nature of the particle. In the simplest case this is obviously true: if one slit is covered, the trajectory of the electron is sure to pass through the other one—but when that is done, the interference pattern, the hallmark of waviness, disappears, in agreement with Bohr's claim. Einstein tried a more subtle approach. In 1927, at a historic international congress in Brussels that was chaired by the venerable Hendrik Lorentz and attended by sixteen other past and future Nobel laureates (including Planck, de Broglie, Bohr, Heisenberg, Schrödinger, and Born), Einstein described a thought experiment designed to contradict Bohr. He imagined that the slits were cut into a movable screen that would receive a little push from every passing electron, to the left if the electron passed through the left slit, to the right if it went through the right one. The motion of the screen would thereby reveal, by the direction of its recoil, which slit had been traversed. (His example actually made use of photons rather than electrons, but the principle is the same.) The result, he claimed, was a counterexample to the Copenhagen interpretation: Even though the electron's wavelike character gives rise to the diffraction pattern, the particle itself must still have a definite trajectory through one single slit. Of course the technology for actually performing such an experiment was utterly inconceivable at the time; the manipulation of single particles lay far in the future.

Bohr swiftly demolished the argument by proving that Einstein's lightweight screen was too delicate. It would be deflected by the passage of an electron so much that the interference fringes would be blurred beyond recognition, like a photograph taken with a shaky camera. Thus, in the case of the movable slit, Bohr was able to pinpoint the specific physical mechanism that safeguards the Copenhagen interpretation, but characteristically he went farther. He argued that the formalism of quantum mechanics has a built-in guardian—Heisenberg's uncertainty principle—that will always keep the electron, or any other particle, from revealing both its particulate and its undulatory nature at the same time.

Forty years later Richard Feynman reiterated the point: "If an apparatus is capable of determining which hole the electron goes through, it *cannot* be so delicate that it does not disturb the [interference] pattern in an essential way. No one has ever found (or even thought of) a way around the uncertainty principle. . . . If a way to beat the uncertainty principle were ever discovered, quantum mechanics would give inconsistent results and would have to be discarded." Categorial pronouncements of this kind are dangerous in science, and have a way of turning out to be overstated.

Bohr won the battle with Einstein at the Brussels conference, but the war continued. The best summary of the debate is an allegory told by Richard Feynman's thesis adviser, John Wheeler. Three baseball umpires are discussing balls and strikes. The first one says, "I call 'em like I see 'em"; the second one, "I call 'em the way they are"; and the third, most reflective one, "They ain't nothin' till I call 'em." If the first attitude reflects a worldview based on the primacy of sense experiences, the second represents Einstein's faith in the existence of an underlying objective reality, and the third one Bohr's philosophy of nature.

The great controversy simmered quietly until 1952, when Einstein's qualitative critique of quantum mechanics was translated into a quantitative theory by the American physicist David Bohm, a student of Robert Oppenheimer. The free-spirited and iconoclastic Bohm was investigated in 1949 by the House Un-American Activities Committee, and upon pleading the Fifth Amendment, he was indicted for contempt of Congress. He was acquitted, but nevertheless lost his job at Princeton University. After an odyssey that took him to Brazil and Israel, he settled in England, where he now lives in retirement. His formulation of quantum mechanics, which differs from the Copenhagen interpretation, resembles an idea de Broglie had abandoned twenty-five years earlier, and is now called the Bohm–de Broglie theory, or, from its principal conceptual ingredient, the guiding-wave theory. At the end of one of his papers Bohm thanked Einstein for "several interesting and stimulating discussions," during which the seeds of the new idea were presumably planted.

The position of the guiding-wave theory in modern physics is curious. The majority of physicists mistakenly believe that it has been proved wrong and dismiss it out of hand. In fact it is to a large extent a reformulation of conventional quantum mechanics and therefore logically equivalent to it, but the two versions differ fundamentally in the interpretation of the symbols they share. Like ordinary quantum mechanics the Bohm–de Broglie theory has some serious shortcomings and has always been severely criticized because of them, even by its defenders. For example, Einstein felt that it was "too cheap," meaning insufficiently profound or novel to be compelling, and physicists generally agree that it is a point of departure rather than a finished edifice. Its principal advantages are that particles are at all times perfectly real, rather than potential, and that it avoids the vexed measurement problem of normal quantum mechanics—two achievements that alone should give the scheme much more prominence than it actually enjoys.

The guiding-wave theory solves the problem of wave-particle duality by meeting it head on: It allows an electron to be a particle whenever it is caught and a wave whenever it passes through two slits. But unlike the Copenhagen interpretation, which insists that the electron is *either* a particle *or* a wave, the guiding-wave theory assumes that it is always a particle *and* a wave.

Specifically Bohm imagined the electron as a particle riding on its wave like a leaf in a brook. This wave, known as the guiding wave, is a real, but invisible, part of every particle. It flows through Young's slits like water and directs the motion of the particle from the source, through one of the slits, to a spot on the screen. Since the guiding wave, which is as real as the gravitational field of the earth and the electric field that surrounds a charge, carries essentially the same information as the totally abstract wave function of conventional quantum theory, the two formulations agree as to their predictions. But in the guiding-wave theory the electron follows a well-defined classical trajectory at all times and never makes quantum jumps from potentiality to actuality.

The Bohm–de Broglie theory has never had many adherents. In the 1960s, when I was a graduate student, Bohm was regarded by many as an eccentric exile, and de Broglie as past his prime. There seemed to be no pressing reason to tinker with the success of quantum mechanics. Then the guiding-wave theory acquired a powerful supporter in the person of John Stewart Bell of the European Center for Nuclear Physics in Geneva, who once described himself as a "quantum engineer" in the same spirit in which the writer Primo Levi called himself a "rigger-chemist"; both men took pride in their mastery of the nuts and bolts of their respective crafts.

Bell's fame rests on a celebrated mathematical theorem, which he proved in 1964 and which now bears his name. Bell's theorem converted Einstein's vague philosophical objections to quantum mechanics into a crisp numerical proposition. It deals with statistical correlations between different particles and shows that quantum mechanical correlations differ from classical ones. Unfortunately statistical correlations are difficult to understand, as the debate about smoking and cancer amply demonstrates. No specific smoker is sure to develop cancer, and no particular cancer victim is sure to be a smoker, yet the strong relationship between the two phenomena is a proven fact. Statistical correlations, in physics as well as in medicine, are hard to grasp because they refuse to talk about individual cases, while at the same time claiming logical rigor and mathematical certainty.

To make matters worse, as if binary correlations—between smoking and cancer, for example—weren't abstruse enough, Bell's theorem concerns three-way correlations that boggle the mind. For example, according to ordinary logic it is an undisputable fact that the number of women with cancer is less than or equal to the number of woman smokers plus the number of nonsmokers of either gender with cancer. Only professional logicians and people addicted to mathematical games find this statement transparent. And yet Bell showed that it is precisely this inequality among three categories that, when applied to the description of the motion of particles, is violated by the rules of quantum mechanics.

In the case of photons, for example, the categories woman, cancer patient, and smoker are replaced by three directions of polarization, and the classically expected inequality becomes a relationship between the number of photons that are able to pass through filters oriented in those directions. The cause of the violation of this inequality is the ability of quantum systems to occupy two contradictory states simultaneously: a photon, for example, can point in two different directions at the same time and thereby defy conventional logic. (A quantum creature could presumably be both male and female, smoker and nonsmoker, cancerous and healthy, all at the same time. Schrödinger pointed out that a quantum cat could even be dead and alive.) In view of the complexity of Bell's theorem it is not surprising that it was at first ignored and later frequently misunderstood by both physicists and the public.

But Bell was a quantum engineer, not a logician, so contact with the real world was of paramount importance to him. Near the end of his brief but revolutionary 1964 paper he wrote, "The example considered above has the advantage that it requires little imagination to envisage the measurements involved actually being made." With these words he challenged his experimental colleagues to devise ways to test his theorem in the laboratory and let nature decide between the validity of ordinary logic on the one hand and quantum mechanics on the other. In due time the experiments were performed, and by the 1980s had decisively reconfirmed the predictions of quantum mechanics. Thanks to his pioneering effort to convert philosophical debates into testable propositions, Bell gained a reputation as one of the most creative contemporary thinkers in the arcane field of foundations (as opposed to applications) of quantum theory.

Bell, who died in 1990 aged sixty-two and at the height of his powers, urged that physics students be taught the guiding-wave theory along with the Copenhagen version. "Despite some curious features," he wrote, "it remains, in my opinion, well worth attention as a model of what might be the logical structure of a quantum mechanics which is not intrinsically inexact." The inexactness he was referring to was the abject failure of quantum mechanics to come to terms with the measurement process, and he felt that this problem is so fundamental that alternative theories, even those that are afflicted with glaring faults, or, as he put it more delicately, "curious features," are worth exploring.

The most curious feature of the guiding wave is what is known as its nonlocality, the manner in which a wave extends far beyond the particle to which it belongs, thus enabling it to affect other particles instantaneously and be affected in turn by distant objects and events. This phenomenon is unknown in the macroscopic world: When a boat on a lake raises a wave, it takes some time for that wave to travel out to another boat, but a guiding wave transmits its influence without time delay. This effect is called action at a distance, and Einstein condemned it as "spooky."

The reason action at a distance is unacceptable to physicists is that they had to work so hard to remove it from the description of nature in the first place. Newton's explanation of gravity was the prototype action-at-a-distance theory. According to this view, any two material objects in the universe attract each other with a force that is transmitted through matter and empty space without any time delay. When a mountain shifts in Australia, or on the moon, the pull of gravity on a mass in New York changes immediately, albeit by an imperceptibly small amount. Action at a distance is magic.

Today physicists regard Newtonian action at a distance as obsolete, at best a crude approximation of the real state of affairs. In truth, gravity, electricity, and magnetism, as well as all other forces, are thought to be transmitted by particles that travel through matter and empty space. Action at a distance corresponds to the assumption that influences spread at infinite speed, whereas in fact Einstein's theory of relativity restricts particles to the speed of light or less. In most modern theories, including the current description of gravity, influences are local: An object is affected only by those force-carrying particles that actually come in direct contact with it. Action at a distance is a relic of a bygone age, and most physicists were profoundly skeptical when the Bohm–de Broglie guiding-wave theory tried to bring it back.

John Bell, on the other hand, proceeded to make a virtue out of necessity. Inasmuch as the guiding-wave theory is equivalent to ordinary quantum mechanics, he argued, and inasmuch as the former is nonlocal, the latter must be too. Bell insisted that the chief advantage of the guiding-wave theory is precisely its power to display the essential nonlocality of the quantum world, its strange feature of action at a distance, which the Copenhagen interpretation manages to hide under a mantle of obfuscation. The absurd nonlocality of quantum mechanics, in either the conventional or the guiding-wave formulation, cries out for explanation, just as the nonlocality of Newton's gravity did. This was the crux of Einstein's disagreement with Bohr, and Bell minced no words about which side he supported: "I feel that Einstein's intellectual superiority over Bohr, in this instance, was enormous, a vast gulf between the man who saw clearly what was needed, and the obscurantist." Bell claimed that the cause of Einstein's objections to conventional quantum theory is brought out into the open by the Bohm–de Broglie interpretation.

In order to learn more about this theory, I went to talk to Jean-Pierre Vigier, a well-known French physicist who is one of its leading exponents. I found him in his tiny cluttered office tucked behind the elevator on the second floor of the Institut Henri Poincaré, on rue Pierre et Marie Curie, in Paris. Vigier was one of de Broglie's favorite assistants, and the long list of his scientific publications includes several he wrote with his illustrious mentor, as well as some with David Bohm. He was seventy-one years old when I met him, but his thatch of silver-streaked hair, his agile square frame, and his quick, forceful way of speaking reminded me of a much younger man. As soon as I had introduced myself and explained what I was after, he pointed to a tiny round armchair whose cracked brown leather and tarnished brass tacks revealed its age more than he did his. "Take de Broglie's chair," he ordered, and squirmed past piles of books around his desk to the blackboard, where he proceeded to launch into a lecture as though I had just come in from the office next door. Caught in the embrace of that low and surprisingly comfortable little chair, feeling the aura of de Broglie suffuse the room, I listened to a report on the status of the guiding-wave theory.

On the theoretical front there is quiet progress. Vigier and a devoted little band of "quantum realists" around the world are extending the Bohm–de Broglie theory by making it consistent with the demands of the theory of relativity, examining a variety of different ideas about the physical nature of the guiding wave, and above all re-deriving those results that conventional quantum mechanics has already successfully explained. The work is difficult and frustrating, because calculations that follow easily from the Copenhagen interpretation are sometimes excruciatingly tedious in the guiding-wave theory. The purpose of this effort is to prepare for the discovery of a significant flaw in the conventional quantum theory, which is a distinct possibility. If such a fault is found, the guiding-wave theorists will be ready to pounce, because at the moment their model is the best alternative to the conventional one. When John Bell surveyed the currently available interpretations of quantum mechanics at a Nobel Symposium in Stockholm in 1986, he wistfully concluded, "In my opinion the [guiding] wave picture undoubtedly shows the best craftsmanship among the pictures we have considered. But is that a virtue in our time?"

Vigier cheerfully admits that his is an uphill battle, but he blames Bohr for his dogmatic attitude. "The Copenhagen spirit is very negative," Vigier told me, "and that's not good for science. Bohr said, 'You can't find out an electron's trajectory, so don't even try.' I think that kind of pronouncement has turned two generations of physicists into mere technicians who just use quantum mechanics and never pause to reflect on its foundations." In a more conciliatory spirit he said, "Maybe that was a good thing, because there was so much work to do. But now it's time to go back and try to do what Bohr declared to be impossible."

None of the theoretical developments excites Vigier nearly

as much as the recent progress in experimental physics. In fact he, along with most adherents of both camps, believes that the Bohr-Einstein controversy is entering a new era. Thanks largely to the insight of John Bell, after sixty-five years of philosophical discussion the debate is finally moving to the only legitimate battlefield for the adjudication of scientific disputes—the experimental laboratory.

Vigier told me about the experiment on which he pins his highest hopes. In the spring of 1991 three scientists at the Max Planck Institute for Quantum Optics in Munich, Marlan Scully, Berthold-Georg Englert, and Herbert Walther, proposed a refinement of the two-slit experiment that aims to accomplish precisely what Einstein had envisioned and Bohr and Feynman believed to be impossible: to circumvent the uncertainty principle. The experiment will involve the manipulation of individual atoms and photons and will therefore be very difficult to perform, but if it succeeds, it will throw new light on the nature of quantum reality.

The double-slit experiment that is being prepared in Munich is basically an example of atom interferometry of the kind demonstrated by Jürgen Mlynek in nearby Konstanz, with one important additional feature. While the atoms display their wave nature by developing a characteristic interference pattern, each one will be tagged with an identifying label that will betray its passage through one slit or the other. Whereas Einstein proposed to tag each particle by its effect on an external, macroscopic device—his movable screen—the new experiment will use internal, microscopic tags: As each atom passes through a slit, it will signal its presence by giving up a photon.

The basic scheme of the experiment will resemble Young's, with the addition, in front of each slit, of a microwave cavity similar to the one used by Daniel Kleppner in his elegant demonstration of the suppression of spontaneous emission. As in that experiment a laser will promote each atom (probably one of rubidium, a metal similar to sodium) to a higher energy state before it enters a cavity. However, where Kleppner adjusted the size of the cavity to minimize spontaneous radiation, here radiation will be maximized instead, so that the vacuum will pull a photon out of each passing atom. Surprisingly, although the atom's electronic structure obviously changes with the emission of a photon, the wave function that describes the atom's motion does not. Thus, by detecting a photon in a cavity, the Munich group has found a way of detecting the atom's position inside a slit without running afoul of the uncertainty principle. The advantage of their scheme over Einstein's jiggling screen is a reflection of the difference between old-fashioned control over macroscopic devices and modern control over individual particles.

The experiment, which, if all goes well, will be performed sometime within the decade, is easily analyzed in terms of conventional quantum mechanics. As long as the microwave cavities in front of the slits are not operating, the atoms will produce the familiar interference pattern that Young discovered for light and Mlynek reproduced two hundred years later with atoms. When the laser and the microwave cavities are switched on, each atom signals the particular slit it has passed through and thereby its trajectory through space and time. According to the Copenhagen interpretation the interference pattern, the fingerprint of waves, will therefore disappear.

This result agrees with Bohr's critique of Einstein's *Gedankenexperiment* in Brussels. However, the absence of an interference pattern can no longer be traced to the uncertainty principle, or to the physical effect of the cavities upon the atoms: Unlike the motion of a particle in Einstein's movable-screen experiment, an atom's motion is unaffected by the emission of a photon in a cavity. The cause of the predicted loss of interference is much more subtle.

According to the analysis offered by the Munich group, the outcome of the experiment is determined by the mere *information* contained in the microwave cavities as the atoms pass. The wave function for the whole system—atoms, photons, and cavities—is one entangled unit. When a photon is detected in one of the cavities, the wave function, which encodes everything that is physically measurable about the atom, including the emission of a photon, carries that information forward to the receiving screen and thereby ensures that the interference pattern is washed out.

To prove that only information, not any physical material, is affected by the interaction of an atom with a cavity, the designers of the experiment have invented one last trick. They propose to build a device called a quantum eraser, which can remove the information gathered by the cavities and cause the interference fringes to reappear.

The quantum eraser is a detector connected to both cavities simultaneously. It will be sensitive enough to extract a single photon from them and will record its arrival without specifying which of the two cavities it came from. In this way the quantum eraser will announce the passage of an atom through the apparatus but at the same time swallow up the clue it left as to its path.

In order to understand the function of the quantum eraser, first consider an experiment in which it remains switched off. An atom passes through the apparatus and gives up a photon, which is stored in one of the two cavities, whereupon the atom creates a spot on the screen. A long time later—perhaps a millisecond, which is an eternity on the atomic scale—the experimenters choose to detect the photon and then proceed to repeat the process with the next atom. Since they have thus discovered the path of each atom, they will find no interference fringes on the screen.

In the second experiment they also let an atom make its way to the screen, but then, instead of detecting each photon, they switch on the eraser before going on to the next atom. Since they thereby lose all information about the path of the atom, they will find that interference fringes do build up. But in both experiments the decision to ascertain the path of the atom, or to activate the eraser instead, is made *after* the atom has already made its mark on the screen.

If the atoms make their marks in exactly the same way in both experiments, how can the screen show fringes in the second case but none in the first? A decade earlier the American physicist Edwin Jaynes had discussed the possibility of just such a quantum eraser: "I say that it constitutes a violent irrationality, that somewhere in this theory the distinction between reality and our knowledge of reality has become lost, and the result has more the character of medieval necromancy than of science."

The Munich group resolves the apparent paradox by showing that while the first experiment proceeds as described, the second one is not as simple as it sounds. Imagine that each experiment is done with one thousand atoms. The initial result of both experiments is the same: When all one thousand atoms have made their marks on the screen, it will show a featureless distribution of dots without fringes—there is no paradox. However, the second experiment, and in particular the eraser, must be analyzed more carefully. Since the eraser detects single photons, it is a quantum device, and obeys quantum logic rather than ordinary common sense. A simple quantum mechanical calculation reveals that half the time it doesn't detect a photon at all, even though an atom has passed through the apparatus. Whereas in the first experiment the cavities would record five hundred passing through the left slit and five hundred through the right one, in the second experiment the eraser would show five hundred going through both slits and five hundred through neither one.

The atoms that fail to trigger the eraser carry excess information, as it were, and must be thrown out of the second experiment. Quantum theory predicts that if, after the conclusion of the second experiment, the spots made by those atoms are removed from the nondescript, smooth pattern on the screen, the remaining five hundred spots will show the characteristic interference fringes. The apparatus seems to be telling us, "When you select those events in which the information gathered by the cavities has definitely been erased, I will reveal the waviness of atoms. But as long as you don't actually erase that information, it will continue to exist somewhere in the system, and I will regard the rubidium atoms as ordinary particles that pass through one cavity or the other, and not as waves."

Before the invention of the quantum eraser the Copenhagen interpretation of the double-slit experiment insisted that when the cavities are inoperative, an atom passing through them is a wave and does not follow a sharp trajectory through them. When the cavities are activated, on the other hand, the theory would say that the atoms are particles with definite trajectories through one or the other. Now these particles can be transformed back into waves and their trajectories retroactively erased from existence. Medieval necromancy indeed.

Professor Vigier did not dispute the predicted outcome of the proposed experiments in his office that day, but offered me a different interpretation from that of the Copenhagen school. In accordance with the Bohm–de Broglie version of quantum mechanics he explained that the atom is always a particle and traverses one cavity or the other, just like a baseball. The interference pattern is accounted for in this view by the guiding wave that accompanies each atom but passes through both slits. However, the details of how that wave is affected by the microwave cavities and the quantum eraser have not yet been fully examined, so it is too early to be certain how the experiment, which in any case hasn't been performed yet, fits into the guiding-wave theory. But Jean-Pierre Vigier is working on it.

As I left his office that day, it seemed to me that a clear-cut victory in the Bohr-Einstein debate may be a long way off. If the Munich experiment succeeds, the adherents of Bohr's view of the world will have scored yet another point, because the outcome will be easily predictable by means of conventional quantum theory. At the same time, Einstein's attempt to beat the uncertainty principle will finally be achieved, giving the idea that atoms are real particles with real trajectories in space and time—as Einstein, Schrödinger, and de Broglie had always hoped—a powerful boost.

While walking home past the shuttered windows of the venerable university buildings along rue Pierre et Marie Curie, I recalled the two pictures that hang over the desk in my study. When they were taken in the early 1920s, Einstein was at the peak of his fame. His theories of special and general relativity had been corroborated experimentally and were accepted by the majority of physicists; both were radical in their underlying assumptions, but both agreed with the realistic world picture of classical mechanics. Bohr, on the other hand, was the harbinger of a new age in which probability would replace certainty, and potentiality substitute for reality. Einstein represented the classical establishment, Bohr the coming quantum revolution.

Today, seventy years later, the roles are reversed. The quantum revolution is long past, and Bohr's Copenhagen interpretation has become established doctrine. Einstein's doubts, on the other hand, now inspire a little band of quantum realists whose aim is to overthrow the current regime—a new revolution is brewing in his name.

In the community of physicists I sense an air of expectancy reminiscent of the mood that must have prevailed in Berlin in the early 1920s just before the advent of quantum mechanics, except for one circumstance. The place of theorists like Einstein and Bohr, who once dominated the debate about the nature of quantum reality, has been assumed by experimentalists who used to remain in the background, alongside my granduncle. The quiet and painstaking efforts of the Munich group, and many others like them, may eventually bring the long-standing controversy to a close; Einstein, for one, was confident that a resolution was possible. His letter to Max Born of September 7, 1944, in which he called his philosophy "antipodean" to that of the Copenhagen school, ends with the words "No doubt the day will come when we will see whose instinctive attitude was the correct one."

14

The Next Revolution

One night in January 1610 Galileo Galilei, then a forty-six-year-old professor of mathematics at the University of Padua, was observing the sky from the garden of the Villa del Selve, twelve miles upriver from Florence. When he turned his newly improved telescope toward the planet Jupiter, he must have been delighted to find its pale disk flanked by four bright little dots glistening like pearls on a stretched necklace. He probably thought they were stars too faint to be seen with the unaided eye and accidentally aligned like the three stars in Orion's Belt, so he sketched their positions and went on to other sights. One can imagine his astonishment when, some nights later, he found the four little dots next to Jupiter again, in spite of the fact that in the meantime the planet had moved a considerable distance through the heavens. The four flecks had evidently followed it, so they were not fixed stars at all, but Jupiter's satellites—Latin for "attendants"—as Johannes Kepler dubbed them a year later. Even more remarkably, their positions with respect to each other and to Jupiter had changed since the first observation. Together with their parent planet they seemed to form a dynamical unit whose parts were mysteriously linked by some unseen agency and whose internal motion demanded to be understood.

An atom in a trap sends a similar message. It blinks to signal that it is not merely an inert particle but a dynamical system, not unlike a planet surrounded by its moons. The atom and the planet are beacons that invite us to explore the strange worlds at the two extreme ends of the scale of distances accessible to the human senses.

Galileo realized the fundamental significance of his discovery as soon as he made it. When, just three months later, he reported his numerous telescopic observations in his first astronomical publication, the *Starry Messenger,* he left the discovery of Jupiter's moons for the end, calling it "the most important in this work." The booklet became a sensation and caused a flurry of telescope manufacture and astronomical observation throughout Europe, though many people rejected its claims. The scientific debate about whether the moons might be optical illusions caused by imperfections in the lenses was an early precursor of the unresolved question of the ambiguities inherent in the images of atoms taken by STM. But in the end the telescope, as simple in design as it has been influential on the course of history, spoke more eloquently than Galileo himself.

One of the principal consequences of that invention was to make the moons of Jupiter resemble our own moon, but the *Starry Messenger* also reported on lunar mountains that resemble terrestrial ones. Thus heavenly bodies gradually ceased being seen as inaccessible bits of heavenly material, as the ancients had believed, and became common objects composed of ordinary rocks and dirt, like the ground we walk on.

This was Galileo's greatest legacy to astronomy. By making the planets accessible—not just to scholars but to all who wanted to see—he bridged the gulf between the terrestrial and the celestial realms that the medieval philosophers had inherited from the Greeks and that had impeded the progress of physics for centuries. A mere ten months after the publication of Galileo's book, the poet John Donne had already understood its true message. Referring to Galileo, he wrote, "Man has weaved out a net, and this net throwne / Upon the Heavens, and now they are his owne." The popular imagination had been stirred, and the scientific revolution was at hand.

The impact of the recent observations of atoms, while not as unexpected as that of Galileo's discoveries, will be just as significant. The mercury atom I saw in its trap in Colorado represents the culmination of the program set in motion twenty-four centuries ago by Democritus. Isolated atoms have an extraordinary emotional appeal for professional scientists, and, having already energized the field of atomic physics to new heights of achievement, promise a variety of practical applications in the future. Of these the most immediate is an improved clock—which also happened to be the first useful result of Galileo's discovery. As soon as he noticed that the moons revolve around Jupiter with steadfast regularity, he suggested that their motion could serve as an astronomical timekeeper, and after his death this idea was implemented.

But just as the philosophical implications of Galileo's discovery overshadowed the importance of its practical applications, the lasting significance of atomic traps goes beyond their scientific and technological benefits. To the world at large the traps will help to transform the atom from a scientific abstraction into a real, material object that is accessible to direct sense experience. They bridge the gulf between the everyday world around us and the theoretical atomic domain first envisioned by the Greek philosophers.

The new three-dimensional color images of atoms made possible by Gerd Binnig and Heinrich Rohrer's STM flesh out what the traps have suggested. We see with our own eyes that atoms are smooth, solid grains of matter, kernels with measurable dimensions and shapes, just as Lucretius had imagined. In the last few years we have even progressed beyond the sense of sight: The magic wrist lets us feel atoms, while other devices allow us to move them about in space, and some version of the tweezers with which Primo Levi dreamed of manipulating chemical structures will soon be listed in scientific-instrument catalogs. Atoms have become real objects, like grains of sand and planets.

They are real enough, but their outward solidity is only apparent. Below the surface an atom has a complex and dynamic inner structure, and its contortions, which have hitherto been too fast to follow, are being charted by Ahmed Zewail's femtosecond lasers. Under the whip of modern technology the ordinary atom is rapidly becoming as tame as Jupiter.

At the same time it is also playing an increasingly significant role in our daily lives. Since time, length, voltage, and electrical resistance are now measured by internationally approved atomic standards, our affairs have become regulated by the properties of atoms in ways we are rarely aware of. Through a convoluted chain of calibrations against the primary standards, a mundane act, such as checking the time or measuring a child's height, represents the encroachment of atomic reality into our everyday world. The conversion of macroscopic standards of measurement into atomic ones has come a long way since it began in the 1960s, and in a few years, when mass, the most corporeal of all physical attributes, is also measured with reference to the atom, the transformation will be complete. The atomic scale will replace the human one as the measure of all things.

This shift is indicative of the future direction of science, which will increasingly turn its attention from human-sized phenomena to the investigation of their underlying atomic structure. Astronomy, for example, has in the past been concerned with understanding the planets and stars in human terms. In antiquity the constellations were identified by their similarities to the figures of the human imagination, and then astronomical distances were related by decreasing stages to human dimensions. Perhaps the clearest remaining expression of the human focus of astronomy is the manned space program, which has dramatically captured the imagination of the public. Scientists tend to argue for robotic space exploration as being cheaper, safer, more versatile, and in the end more illuminating. The question, they claim, is not so much how Martian soil feels to the touch as the nature of its chemical composition, and modern astronomy has gradually become more preoccupied with the atomic nature of the universe than with how it reflects the human experience.

In the microscopic realm the change of emphasis is also well under way. Biology, which used to be solely concerned with the classification of plants and animals into man-made categories, has given rise to the fields of biophysics and biochemistry, which attempt to explain living organism in terms of atomic processes. Modern chemistry deals less with vats and retorts than with mass spectrometers and lasers, and physics itself has turned its attention from the mechanics of falling apples and flying cannonballs to the motion of atoms. Eventually the fulcrum of all scientific activity will be the universal building block of matter.

In spite of its growing importance, however, the atom's inner workings continue to elude common sense. In this respect, too, the atom resembles the Jovian satellites, except that visual observation and theoretical explanation occurred in opposite order for the two phenomena. The discovery of Jupiter's moons preceded the description of their orbits in terms of Newton's theory of gravity by a couple of generations, whereas pictures of atoms followed the quantum theory by about the same interval. But aside from this historical asymmetry there remains a striking parallel between the two mechanisms.

When Isaac Newton formulated the law of universal gravitation in 1666, he immediately applied it to the motion of the Moon, the orbits of the planets, and the revolution of the satellites of Jupiter. Since it agreed in minute detail with all the astronomical observations of the time, it grew in stature until it became the sturdiest pillar of classical mechanics. Schoolchildren learn this law and cite it as the obvious reason for the fall of an apple from a tree and the journey of the Moon through the sky.

But of course it is not obvious—it is not even reasonable. The notion that two distant objects should exert a pull on each other, as if they sprouted unseen tentacles that could reach across empty space to tug them into a tighter embrace, is fantastic; nothing in our daily experience suggests such an effect. In order to bring a sugar bowl toward me, I must touch it, and in order to send it back, I must push it away. Billiard balls do not deflect each other except by immediate touch, and even the force of the wind is mediated by particles of air that physically touch the trees they bend. Local action conforms with common sense; action at a distance is sorcery.

No one understood this better than Newton himself. In 1693

he wrote four letters to the brilliant young clergyman Richard Bentley, who had requested help in understanding the Newtonian philosophy. On the subject of universal gravitation he remarked, "That gravity should be innate, inherent, and essential to matter so that one body may act upon another at a distance through a vacuum without the mediation of anything else by and through which their action may be conveyed from one to another is to me so great an absurdity that I believe that no man who has in philosophical matters any competent faculty of thinking can ever fall into it."

Newton realized that his law provided an accurate description of the phenomena but lacked explanatory power. And yet, with our own eyes we see the curious effect of Jupiter on its moons, and we teach our children that the law of gravity explains how it works.

Newton was not alone in perceiving the inadequacy of the idea of action at a distance; other clear thinkers also distrusted it. René Descartes, who died when Newton was a boy and who helped to lay the philosophical foundations of the scientific revolution, proposed a theory in which gravity was mediated by whirlpools of invisible particles that permeated all space. Not satisfied with his own theory, Newton took this hypothesis sufficiently seriously to examine and refute it in detail, but his followers were less thoughtful. Dazzled by the practical triumphs of the universal law of gravity, they elevated it to the status of gospel. For a full 250 years the law that its own creator regarded as absurd was praised as one of the grandest conceptions of the human mind and cited as the paradigm of the scientific method. It reigned from 1666 until 1916, when Albert Einstein explained and corrected it by showing that its mediating agent is space itself, and thereby replaced the concept of action at a distance with the far more natural notion of local action.

Many of the architects of atomic theory, including Planck, Einstein, de Broglie, and Schrödinger, were no less troubled by the interpretation of quantum mechanics than Newton was by gravity. Einstein's characterization of quantum mechanical action at a distance as "spooky" perfectly echoes Newton's description of gravity as an "absurdity," and the modern reaction to quantum mechanics bears a striking resemblance to the enthusiastic response of the eighteenth and nineteenth centuries to the Newtonian doctrine. Until recently physicists were so busy working out the rich implications of the theory that they had no time to ponder its foundations. For the most part they accepted quantum mechanics for what it was worth and rejoiced that they had been privileged to penetrate so far into the secrets of the universe. They were grateful for the glimpse under the great veil that quantum mechanics allowed, and accepted its strangeness as the unavoidable price of scientific progress.

But that attitude is changing. "Nobody understands quantum mechanics," Richard Feynman complained; "necromancy" was the word Edwin Jaynes used to characterize the quantum eraser; "bizarre" is how John Stewart Bell described the action-at-a-distance aspects of the theory. Such phrases reveal the undercurrent of skepticism that is once more bringing the meaning of quantum mechanics to the forefront of research.

The two rival conceptions of the atom are coming into focus with increasing clarity: The ordinary atom, which is corporeal and substantial, a thing no different from the great Jovian atom in the sky, is taken for granted as a constituent of all matter, while at the same time the scientific atom, which is described in the alien language of quantum mechanics, remains a theoretical abstraction. The probability interpretation of the wave function, the nonlocality of the theory, the enigmas of quantum jumps and the act of measurement, the uncertainty principle, the duality of waves and particles, and the exclusion principle that keeps electrons away from each other—all of these complexities render the interior of the atom unimaginable.

We can no longer ignore the contradiction between the two conceptions; the extraordinary technical advances of the last decade will compel us to face it head on. As quantum mechanical phenomena are lifted up into the macroscopic world by experiments like Claudia Tesche's forthcoming investigation of supercurrents, wave functions are molded into particlelike shapes by Carlos Stroud's laser pulses, and atomic interferometry reveals quantum mechanics in action, the ordinary atom will be forced to confront its scientific doppelgänger.

Both perceptions of the atom—the miniature grain of sand and the quantum mechanical phantom—exist without a shadow of a doubt. The first we see and the second we know, and the gulf between them is profound. The question is how will we deal with it.

There are two possibilities: We either live with it or we find a way to bridge it. If quantum mechanics emerges as the only true picture of reality, the human intuition will gradually have to learn to adapt to the proposition that in principle a skier really can pass on both sides of a tree, that a car can tunnel through the wall of a garage, and that a baseball can occupy two places at once. Probability and potentiality would become more natural languages for describing the world than strict determinism, and the law of the excluded middle, according to which two contradictory statements cannot hold simultaneously, would be discarded. All things would be connected to each other, and relationships among objects would turn out to be just as fundamental as the objects themselves. The way humans perceive physical reality would differ from today's perception as profoundly as today's materialistic perspective differs from the medieval spiritual one.

If quantum mechanics were to win, the ordinary atom, and with it the ordinary world around us, would be reduced to the status of illusion, or, at best, approximation to reality. Physicists would have to confront the question of how a dreamlike world that appears so real to our senses could emerge out of the underlying quantum mechanical truth. How solidity could result from fluidity, how continuity could come from granularity, how certainty could arise from probability, and how substantiality is mimicked by its very opposite—all these would have to be explained.

The ascendancy of quantum mechanics, which would represent a victory for Bohr's point of view over Einstein's, would spell an end to the dominance of the doctrine of atomism. Atomism, which is based on the faith that physical phenomena can be understood in terms of a finite number of ultimate building blocks and their interactions, has been spectacularly successful in the study of matter, but quantum mechanics conflicts with it in a fundamental sense. The quantum mechanical point of view treats atomic particles as potential, rather than actual, objects whose realization depends on the details of the external, macroscopic measuring devices that are used to detect them; that is, on their physical surroundings. Some writers claim that quantum mechanics blurs the boundary line between the observer and the observed, but it might be more accurate to substitute the word *environment* for the term *observer*. In any case quantum mechanics denies the possibility of analyzing the world in terms of its irreducible components and thereby violates the spirit of atomism.

But there is another way out of the dilemma: Perhaps the gulf between the ordinary and theoretical versions of the atom will be abolished in a second quantum revolution. No scientific theory lives forever; on the contrary, the ultimate vindication of a theory is to become embedded in a more general framework. Newton's law of gravity survives as an approximation of Einstein's general relativity; Maxwell's nineteenth-century equations have become axioms of modern quantum electrodynamics; and even Greek atomism, modified by the demands of quantum mechanics, remains an underlying theme of today's particle physics. In the same way quantum mechanics may one day emerge as part of a grander scheme.

From what quarter the revolution will come is impossible to predict, but it will most likely begin with a deficiency in the present atomic theory, where it is flawed or incomplete, such as its intrinsic nonlocality, its seeming incompatibility with the theory of gravity, and its failure to describe the measurement act. In these three areas scrutiny of the theory is most intense, and the search for alternatives most urgent. Any of these difficulties may be turned to advantage if they provide clues to a more comprehensive theory and thus a more perfect understanding of the atom, and all three have inspired speculations, without giving definitive answers.

It is conceivable that a more general theory would turn out to be so complex that it would drown the mysteries of quantum mechanics in a computational morass. Such intricacies exist in the Bohm–de Broglie theory, for example, in which the wave function for a double-slit interference experiment resembles the jumbled topography of the Himalayan mountain range. In such a theory the atom assumes a kind of mediated reality, like the unimaginable turbulence of the interior of the sun, which can be described only by a supercomputer: We understand the basic equations, but their meaning for any specific case can only be discovered by means of a calculating machine.

The eminent American theoretical physicist John Wheeler, who supervised Richard Feynman's thesis, and later explained the Bohr-Einstein debate in the language of baseball umpires, is much more optimistic. He refuses to hide behind the fashionable word *complexity* and goes for broke:

> If one really understood the central point [of quantum mechanics] and its necessity in the construction of the world, one ought to be able to state it in one clear, simple, sentence. Until we see the quantum principle with this simplicity we can well believe that we do not know the first thing about the universe, about ourselves, and about our place in the universe.

And then, at a conference in 1984, he made a remarkable prediction that could serve as inspiration for all who want to understand quantum theory: "The most revolutionary discovery in science is yet to come! And come, not by questioning the quantum, but by uncovering that utterly simple idea that demands the quantum."

That momentous discovery will require taming the atom, for, as the fox tells Saint-Exupéry's little prince, "One only understands the things that one tames." And when the little prince asks how to go about the business of taming, the fox replies: "You must be very patient. First you will sit down a little distance from me—like that—in the grass. I shall look at you out of the corner of my eye, and you will say nothing. Words are the source of misunderstandings. But you will sit a little closer to me, every day . . ."

The myth of *The Little Prince* speaks to us about the method of physics with greater immediacy than any scholarly treatise. It reminds us that observation is the final arbiter of scientific truth, and the only reliable guide past the misunderstandings that abstract theory can lead us into. After more than two thousand years of hearing words about the atom, we have finally learned to see it, first out of the corner of our eye, and then coming closer every day. The bond of understanding we are thereby establishing with the atom will endow it with deeper meaning, until one day a profound and simple idea will resolve the enigma of the quantum.

When the time came for the fox to leave, he offered the little prince a parting gift: "Here is my secret," he said, "a very simple secret: It is only with the heart that one can see rightly; what is essential is invisible to the eye." "What is essential is invisible to the eye," the little prince repeated, so that he would be sure to remember it. In physics, as in human affairs, the deepest truths cannot be seen—they must be felt. The great physicists knew that: When Isaac Newton discovered the law of gravity, he knew in his bones that it was absurd; young Werner Heisenberg trusted his intuition to assure him that his strange new theory was sound; and Albert Einstein, who understood the mathematical formulation of quantum mechanics as well as anyone, and could see its marvelous successes, nevertheless felt deeply that we had not yet grasped the atom's invisible essence. When we finally do, the way we perceive the world will undergo a radical change.

If the analogy between the atom and the planet Jupiter is a guide, we may have to wait a quarter of a millennium for this revolution, during which time quantum mechanics will have to be accepted as uncritically as Newton's law of gravity was. On the other hand, twentieth-century physics furnishes a more auspicious historical precedent. Quantum mechanics began in the year 1900 when Max Planck, while trying to patch a seemingly insignificant crack in the foundations of classical mechanics, stumbled upon the first piece of the theory. Since then quantum physics has been built into a sturdy edifice, while its foundations have remained underground, inaccessible to direct experimentation. Now they have been unearthed; the atom has become visible and tangible, quantum effects have been raised from the atom to the laboratory, yesterday's *Gedankenexperiments* have become real experiments, and industrial physicists are beginning to exploit the strange behavior of atomic systems in unusual new devices. These foundations, which have been left virtually unexamined for fifty years, are once again becoming a lively topic of research, and the atom itself has become the principal tool in this pursuit. The circumstances are propitious for the third millennium to begin, like the twentieth century, with a wonderful new insight into nature's grand scheme.

And then, even as the atom emerges into public view like a jewel of exquisite design, it will open up and allow, for the first time since the dawn of science, a glimpse inside.

About the Type

This book was set in Century Schoolbook, a member of the Century family of typefaces. It was designed in the 1890s by Theodore Low DeVinne of the American Type Founders Company, in collaboration with Linn Boyd Benton. It was one of the earliest types designed for a specific purpose, the *Century* magazine, because it was able to maintain the economies of a narrower typeface while using stronger serifs and thickened verticals.